Urban Water Sustainability

The provision of a safe and reliable water supply is a major challenge for the world's growing urban populations. This book investigates the implications of different developments in water technology and infrastructure for urban sustainability and the relationship between cities and nature.

The book begins by outlining five frameworks for analysing water technologies and systems – sustainable development, ecological modernisation, socio-technical systems, political ecology and radical ecology. It then analyses in detail what the sustainability implications are of different technical developments in water systems, specifically demand management, sanitation, urban drainage, water reuse and desalination. The main purpose of the book is to draw out the social, political and ethical implications of technical changes that are occurring in urban water systems around the world, with positive and negative impacts on sustainability.

Distinguished from existing social science analysis due to its attention to the engineering details of the technology, this book will be of use to a wide audience, including students on water management courses, engineering students and researchers, urban geographers, and planners interested in sustainability, infrastructure and critical ecology.

Sarah Bell is Professor of Environmental Engineering and EPSRC Living with Environmental Change Research Fellow in the Bartlett School of Environment, Energy and Resources at UCL, UK. She is also Director of the Engineering Exchange, which serves to improve community engagement with engineering and built environment research. She is a Chartered Engineer and Fellow of the Chartered Institution of Water and Environmental Management.

Earthscan Studies in Water Resource Management

For more information and to view forthcoming titles in this series, please visit the Routledge website: www.routledge.com/books/series/ECWRM/

Water Policy, Imagination and Innovation
Interdisciplinary Approaches
Edited by Robyn Bartel, Louise Noble, Jacqueline Williams and Stephen Harris

Rivers and Society
Landscapes, Governance and Livelihoods
Edited by Malcolm Cooper, Abhik Chakraborty and Shamik Chakraborty

Transboundary Water Governance and International Actors in South Asia
The Ganges-Brahmaputra-Meghna Basin
Paula Hanasz

The Grand Ethiopian Renaissance Dam and the Nile Basin
Implications for transboundary water cooperation
Edited by Zeray Yihdego, Alistair Rieu-Clarke and Ana Elisa Cascão

Freshwater Ecosystems in Protected Areas
Conservation and management
Edited by Max C. Finlayson, Jamie Pittock and Angela Arthington

Participation for Effective Environmental Governance
Evidence from European Water Framework Directive Implementation
Edited by Elisa Kochskämper, Edward Challies, Nicolas W. Jager and Jens Newig

China's International Transboundary Rivers
Lei Xie and Jia Shaofeng

Urban Water Sustainability
Constructing Infrastructure for Cities and Nature
Sarah Bell

Urban Water Sustainability

Constructing Infrastructure
for Cities and Nature

Sarah Bell

Routledge
Taylor & Francis Group

LONDON AND NEW YORK

First published 2018
by Routledge

2 Park Square, Milton Park, Abingdon, Oxfordshire OX14 4RN

52 Vanderbilt Avenue, New York, NY 10017

Routledge is an imprint of the Taylor & Francis Group, an informa business

First issued in paperback 2020

British Library Cataloguing-in-Publication Data
A catalogue record for this book is available from the British Library

Library of Congress Cataloging-in-Publication Data
A catalog record for this book has been requested

ISBN: 978-1-138-92990-6 (hbk)
ISBN: 978-0-367-59348-3 (pbk)

Typeset in Sabon
by Apex CoVantage, LLC

Contents

List of figures	vi
List of tables	vii
List of boxes	viii
Preface	ix
Acknowledgements	xii
1 Introduction	1
2 Water and sustainable cities	12
3 Constructing infrastructure	24
4 Framing cities and nature	35
5 Demand	56
6 Sanitation	84
7 Drainage	106
8 Reuse	131
9 Desalination	152
10 Conclusion	174
Index	181

Figures

5.1   Water consumption in selected cities   58
5.2   International comparison of micro-component water
      demand   62
6.1   Activated sludge process   88
6.2   Urine diverting toilets, eThekwini Municipality, South Africa   93
6.3   Loowatt waterless, container-based toilet in Madagascar   95
7.1   Natural, conventional and sustainable drainage   108
7.2   Urban stormwater hydrographs: pre- and post-development   109
7.3   The Thames Barrier   110
7.4   Combined sewer overflows   113
7.5   Green roof with solar cells and food growing in Portland,
      Oregon   115
7.6   Rain garden receiving runoff from pavement and road
      in Portland, Oregon   116
9.1   Multi-effect distillation desalination   155
9.2   Multi-stage flash desalination   156
9.3   Reverse osmosis desalination plant   158

Tables

4.1 Frameworks for urban water sustainability   51
5.1 Factors shaping water demand   61
5.2 Theories and methods for reducing water demand   65
6.1 The sanitation ladder   96
9.1 Energy intensity of water treatment technologies   153

Boxes

5.1   New York City water demand management   67
7.1   Philadelphia Green City, Clean Waters   120
7.2   China's Sponge City programme   122
8.1   Potable reuse referendum in Toowoomba, Australia   135
9.1   Desalination in Chennai   162

Preface

As I write this book in the summer of 2017, more than half a million people have contracted cholera in Yemen, and nearly 2,000 have died. Drinking fountains in Rome are being turned off and nearby towns face water rationing as drought continues and more than a third of water supplied is lost from old, leaky pipes. Here in London, the last cholera epidemic was in 1866, just prior to the completion of the intercepting sewer network, one of the biggest infrastructure projects this city has ever seen. Since 2012, London has had a desalination plant sitting idle at great expense, just in case of severe drought, while leakage remains around 25%.

Water infrastructure matters. It protects cities from disease and provides a buffer against seasonal variability in rainfall and the extremes of drought. Water infrastructure is never finished. While Londoners now live without fear of cholera, supplying water to a growing population with finite water resources, fixing leaks, preventing overflows of sewage from nineteenth-century sewers into the Thames, and paying for a desalination plant that has never been used are just some of the complex challenges facing water managers and citizens. Water infrastructure is even more of a challenge in cities of the Global South, which are still struggling to deliver universal access, good public health and basic environmental protection.

Water infrastructure and sustainability defy categorisation into the boxes of technology and society that have come to define our universities, professions and institutions. Technology, society, politics and ecology get all mixed up in trying to talk about sustainable urban water systems. Interdisciplinary approaches are needed, but interdisciplinarity is hard. I've been trying for decades.

I started my career as an environmental engineer in the 1990s. At the University of Western Australia, I studied flow in pipes and channels, rainfall-runoff hydrographs, water and wastewater treatment, the hydrodynamics of lakes and rivers, and the impact of pollutants on human health and ecosystems. In an 'engineering communications' course we debated whether sustainable development was an oxymoron, and I took the side of the optimists – surely development was possible without destroying the planet.

I learned that technology could solve problems, but as a graduate engineer in the aluminium sector I also learned that technology choices are constrained by economics, regulation and the sheer momentum of industrial growth. I wrote an essay for a part-time master's course at the University of New England about the ethics of climate change, and I decided to go back to university to figure out why our technologies were still so bad for the environment. If engineers could solve most technical problems they were set, it had to have something to do with society, politics and economics.

I started a PhD at the Institute for Sustainability and Technology Policy, then at Murdoch University in Perth. There I met scholars who questioned the fundamental basis of human relations with the natural world. I read deep ecology and ecofeminism, and I talked to farmers about the Oil Mallee Project, an initiative to reintroduce indigenous trees back into agricultural landscapes buffeted by global trade and local hydrology. Theories about the dangers of the domination of nature by Western industrial culture rang true as I drove around wheat fields threatened by land salinisation and visited rural towns that were rapidly depopulating, but they weren't much help in figuring out what to do about it. I discovered actor-network theory and ideas that interpreted the modern world as a series of interconnected socio-technical systems. I used them to structure my PhD dissertation and to think about sustainability as an interdisciplinary endeavour.

Sometime later I landed in London, and I returned to water and engineering. At UCL I worked with urban geographers and planners who pointed out that water and power often flow in the same direction. They showed me that the fact that the poorest, most marginalised people in cities are also the most vulnerable to water infrastructure failure is not a coincidence, but the outcome of a form of politics dominated by neoliberalism and global capitalism. I also worked with engineers who remained convinced that technological and economic efficiency were the key to sustainability, and ecologists who turned their scientific understanding of the natural world into 'ecosystems services', so that economists could value it and policy-makers take notice.

Eventually, I came around to writing this book. I wanted to analyse the technologies of urban water infrastructure in a way that represented the complex relationships between cities, nature and water. But who was right? The teenager in the 1990s excited by the possibilities of sustainable development? The young engineer working to design technologies to protect the environment and make money? The wide-eyed PhD student captivated by philosophies of living in harmony with nature? The researcher putting methodologies from science and technology studies to work in understanding sustainability? Or the global citizen worried about the unequal distribution of wealth and environmental costs and benefits?

My own journey as a researcher and engineer reflects wider debates about sustainability, technology, society and nature. The stages of my interdisciplinary development and the colleagues I met along the way map onto wider theoretical and political frameworks – sustainable development, ecological modernisation, radical ecology, socio-technical systems and political ecology. Each of these positions gave me useful insights, but none of them had all the answers. And so each has its place in my book about water in cities, the subject I have spent the last 12 years researching.

After many years of trying to choose a theoretical framework, or join an academic tribe, I have come to value pluralism. Each of the perspectives I've explored across my career and in the research for this book reflects different politics and values, ways of understanding the world as it is and deciding how it should be. I can analyse the merits and consequences of different positions, but ultimately it is not for me as a researcher to decide which is 'right'.

As global and national politics seem to be moving out of a period where one particular form of politics, neoliberalism, has dominated public policy and discourse, pluralism offers hope. It is possible that the world will emerge into another dominant form of politics that will structure relationships with each other and the natural world for decades to come, but it is not possible at this point to say what that will be, and it certainly will not be decided by academics working in universities.

And so it is with urban water sustainability. There are many different ways of understanding water in cities and working towards a sustainable future. My intention here is to make them apparent to others working and studying in this field, to provide some structure to the rowdiness of debate about how to build cities where people can live healthy, fulfilling lives without undermining the hydrological and ecological systems that we are part of.

This book is not about me. It's about water in cities. I have written it in the absent third person, the standard voice of authority, but my own values and experiences shape the text as much as the data and research it refers to. This story of my own meandering career and academic exploration might help you to read the book as an intellectual and personal journey as well as an analysis of urban water sustainability and contribution to deliberation about the kinds of cities we want to live in.

Sarah Bell, August 2017

Acknowledgements

This book is the outcome of more than a decade of research and more than a year dedicated to writing. Many people and institutions contributed to the development of the ideas and provided the space and time to get them onto paper.

The Engineering and Physical Sciences Research Council funded half of my time since 2016 through a Living with Environmental Change Research Fellowship, which provided the flexibility to be able to dedicate time to research and writing.

A Dyason Fellowship from the University of Melbourne allowed me to work with Anna Hurlimann in the Melbourne School of Design. Our conversations were a great backdrop for beginning of the end of the research for this book.

Colleagues in the Department of Civil, Environmental and Geomatic Engineering at UCL worked around my research and writing commitments, allowing me to disappear for long periods. They also provided an environment where thinking strange interdisciplinary thoughts and taking on ambitious intellectual projects was considered quite normal, for which I am grateful. The Institute for Environmental Design and Engineering provided a quiet desk for finishing the project, and a new home for future research. The UCL Grand Challenge for Sustainable Cities, Environment Domain and the Urban Laboratory helped me to learn about cities, development and interdisciplinarity from very clever and generous colleagues from all over the university.

These ideas have been developed in collaboration with many outstanding students who have worked with me over the years. Joseph Hillier, Kate Crawford, Andrew Chilvers, Tse-Hui Teh, Tori Aitken, Ashley Woods, Luke Mitchell, Crystal Fenwick, Vera Bukachi, Atiyeh Ardakanian, Pascale Hofmann, Fernanda Garcia Alba Garcidiego, Usman Akeel and Zeyu Yao all challenged and inspired my thinking about water, cities, technology and sustainability as I supervised their research degrees. Many master's and undergraduate projects are the source of much of what I have learned and written down on these pages. The international ambitions of this book reflect the diversity of students I've taught at UCL. Thanks to all who told stories and showed me around the cafes and drains of their hometowns.

As a researcher, I am lucky to have been able to work with people who are actually doing urban water sustainability. Amongst them are Martin Shouler, Alexa Bruce and Siraj Tahir from Arup; Abby Crisostomo from KLH Sustainability; Aaron Burton, Jacob Tompkins, Ike Omambala and colleagues from Waterwise; Sian Hills, Matthew Greetham, Paul Rutter, Marie Raffin, Andrew Tucker and others from Thames Water; Paul Schaffer at CIRIA; and Samantha Heath from the London Sustainability Exchange.

Academic colleagues in the UK and international community of water researchers have encouraged me to test ideas and push interdisciplinary boundaries. Some of them are Liz Sharp and Richard Ashley from the University of Sheffield; Ana Mijic at Imperial College London; Catharina Landstrom from the University of Oxford; Paul Jeffrey and others at Cranfield University; Ali Browne from the University of Manchester; Zoe Sofoulis from the University of Western Sydney; everyone else in the WATEF Network; and all the gang at the University of Exeter Centre for Water Systems.

Working with David Goldberg of ThreeJoy Associates helped clarify my career priorities and take the plunge in kicking this whole thing off.

The loneliness of writing was made bearable by endlessly accommodating and amusing friends and family. Thanks to Brigid, who once again gave me her spare room to work on my latest project. Lisa, Louise, Katie, Mum, Dad, Pat and the others let me bring my laptop on holidays, and occupied themselves and did the housework while I was stuck at the computer. Barbara, Paul and Sophia provided a rare mix of intellectual rigour, good cheer and neighbourliness. And to thanks everyone else who looked bemused as 'still writing' became the standard answer to any enquiry about my wellbeing.

The ideas presented in this book were developed through two journal articles. I am grateful for feedback from editors and reviewers, and for permission to reprint extracts from:

Bell, S., 2015. Renegotiating Urban Water. *Progress in Planning* 96, 1–28.
Bell, S., forthcoming. Frameworks for Urban Water Sustainability. *WIREs Water*.

Adam Walls from the Bartlett School of Architecture drew the illustrations with short notice and great professionalism and skill.

Loowatt provided permission to reprint the image of their toilet in Figure 6.3.

Ashley, Tim and Amy at Earthscan-Routledge were very patient and professional in working with me to make this book real.

1 Introduction

Urban water sustainability

Water has always been an active constituent of cities. It is a basic requirement for human health and wellbeing, a source of risk through flooding
and contamination, a remnant of natural hydrology and ecology in streams
and wetlands, a transport route along rivers and canals, and an element of
urban design in fountains, ponds and water features. Water infrastructure
allows cities to function safely within upper and lower hydrological limits –
providing a constant supply of water during dry seasons and droughts, and
preventing flooding during high rainfall events. It allows urban citizens to
go about their daily lives without being preoccupied with where their water
is coming from or going to. Sustainable urban water systems aim to achieve
this within environmental limits to water, energy and pollution, in ways that
are affordable and equitable and contribute to ecological restoration.

Urban water sustainability aims to manage water in cities to provide
for human health and wellbeing within hydrological and ecological limits.
Urban water systems include drinking water supply, wastewater disposal,
surface water drains and the rivers, streams, wetlands and aquifers of urban
water catchments. Urban water sustainability is presented under different
labels, for example, sustainable urban water management (SUWM), integrated urban water management (IUWM), and water sensitive cities. As a
progressive movement, it anticipates positive change in cities. Sustainable
urban water systems are commonly represented as future urban water systems, developed in response to resource constraints, growing populations
and climate change.

The sustainability of urban water infrastructure must account for the relationship between the city and its hydrological catchments. Urban water use
and pollution have impacts in catchments beyond the city limits. As infrastructures expand to meet growing demand, catchments for urban water
supply and waste discharge do not necessarily conform to the geographical boundaries of river basins. Cities draw on regional water resources and
some use water transferred over long distances. Urban wastewater and runoff pollutes rivers, estuaries and coastal environments. The energy used for

pumping and treatment of water and wastewater also creates wider impacts and demands on the environment. How water is managed in cities affects health, wellbeing and the environment locally, regionally and globally.

Urban water crises?

Consistent with global environmental discourse, professional and academic movements in support of urban water sustainability are typically founded on the premise that current water systems are unsustainable. The case for the unsustainability of existing infrastructure is made in terms of limits to freshwater resources, environmental impacts of water abstraction and pollution, growing demand for water due to population growth, rising costs of infrastructure provision, replacement and expansion, vulnerability to droughts and floods, and climate change (Marlow et al., 2013; Mitchell, 2006; Niemczynowicz, 1999; Novotny et al., 2010). Sustainability is presented as the means to avoid crises of water shortages and floods in cities, and the related concept of resilience enables sustainable cities to respond more effectively to extreme events, such as drought and flood, which are more likely in an uncertain future (International Water Association, 2015).

Water scarcity provides some of the most alarming projections and warnings about climate change, population growth and development in popular and policy discourse. However, increasing water scarcity in most regions will be driven by increasing demand for water, rather than decreasing supply due to climate change, and the experience of insecure water supply for the world's poorest people is most often the result of political and economic failure, rather than hydrological constraints (Arnell, 2004; Zeitoun et al., 2016).

Water is important for sustainable cities, but cities have a relatively small direct impact on the sustainability of global water resources. Globally, around 70% of water is used for agriculture, 10% for municipal supply and the rest for industry (Oki and Kanae, 2006; Shiklomanov, 2000). Global and regional forecasts of water scarcity mostly result from unsustainable abstraction for agriculture. Given that more than 50% of the global population live in cities, urban food consumption rather than urban water use is the most significant factor in achieving sustainable water resources management at the global scale (Allan, 2011).

Urban water sustainability is a global goal for development and environmental protection, but it is experienced in localised contexts under conditions of inherent uncertainty. Global and regional assessments of water resource availability can hide specific local conditions. Whilst municipal water supply may not constitute the largest proportion of global water resource use, an individual city can have significant impact on the hydrology of its local catchment (United Nations Human Settlements Programme, 2006). Local hydrology, ecology, urban form, governance, climate, economics, society and other factors shape the form of urban water infrastructure and responses to problems of water scarcity, pollution, flooding and access to water and sanitation services. Cities in the Global South may be focussing on provision of water and sanitation services to a rapidly growing population, while cities with established infrastructure focus on reducing demand and pollution, and restoring degraded freshwater ecosystems (Russo et al., 2014; UN-HABITAT, 2003; Wong and Brown, 2009).

It is common in urban water sustainability to speak of the need for a paradigm shift. Authors including Brown et al. (2009), Novotny et al. (2010) and Allan (2006, 2005) position an emerging paradigm of sustainability and integrated water management in the context of a longer history of water infrastructure. Brown et al. (2009) analyse the history of urban water management in Australian cities and identify six regimes, beginning with early European settlement and projecting into the future. The regimes are the water supply city; the sewered city; the drained city; the waterways city; the water cycle city; and the water sensitive city. Novotny et al. (2010) identify four historical paradigms from ancient times to the modern era: basic water supply; engineered water supply and runoff conveyance; fast conveyance with no treatment; and fast conveyance with end of pipe treatment. They make the case that a fifth emerging paradigm of sustainability will lead to the creation of water-centric ecocities. Tony Allan (2006, 2005) analyses the history of hydro-politics to identify five paradigms of water management – pre-modern; industrial modernity and the 'hydraulic mission' of large-scale infrastructure; environmental awareness; the economic value of water; and integrated water resources management (IWRM), which is still emergent. Allan's analysis recognises paradigms as policy discourses, which may co-exist and contradict each other, in contrast to more linear, progressive notions of paradigms inevitably leading towards sustainability.

The global discourse of environmental crisis that underpins much of the justification for sustainable development and the need for new paradigms of urban water management sits somewhat uncomfortably with the experience of water infrastructure in most cities. Cities can have significant impacts on local water resources and ecosystems, but lack of access to water in cities is rarely the result of water scarcity. Inadequate water and sanitation is usually due to lack of infrastructure, not lack of water (Bakker, 2010; Cook and Bakker, 2012; Zeitoun et al., 2016). With notable exceptions in recent years, few cities in the world face absolute water scarcity sufficient to risk public health. Infrastructure management is typically a bigger factor in water shortages than lack of resources.

Providing water and sanitation to a growing population is undoubtedly challenging, and will impact local resources and ecosystems, but water resource constraints are not the only, or even the most important, factor driving the move towards more sustainable water systems. Beyond alarmist calls to avert catastrophic collapse, urban water sustainability provides an opportunity to reconsider how cities relate to water resources and the natural environment. Water systems reflect wider sustainability challenges of reducing consumption and pollution, improving equality of access and providing a safe environment in which people and nature can thrive.

Constructing infrastructure

Urban water infrastructures, whether sustainable or not, are human inventions. The particular form of an urban water system is an outcome of design and decision-making. Infrastructure is more than pipes, pumps and treatment works. Infrastructure systems cannot operate without ongoing administrative and institutional structures. They include systems of management, regulation, governance, finance and expertise. Water infrastructures do not exist in technical isolation, but are always deployed within political and social contexts, and are shaped by different knowledge and values. Just as the form of pipes and treatment vary in different cities around the world and at times in history, so the forms of governance and administration of infrastructure also change. Infrastructure is therefore constructed – physically and socially.

The form of water infrastructure shapes and responds to daily life in cities. Water infrastructure that was intended to improve public health has evolved to meet demand for water for automatic washing machines, dishwashers, showers, swimming pools, car washing, lush lawns and candlelit, scented bathtubs (Shove, 2004). The provision of constant water supply and wastewater disposal, as well as reliable drainage networks, have opened up new ways of living in cities. Infrastructure and society are constantly co-evolving in cities, with important implications for resource demands and ecological integrity.

In order to understand the role of technology and infrastructure in sustainable cities it is therefore important to be able to discern how their meanings are constructed within different cultural narratives and political discourses, as well as to understand their technical and physical performance. This is particularly important given the long-term nature of sustainability. Political and cultural discourse can change rapidly, while infrastructure may last for centuries and hydrological systems adapt and evolve over millennia. Understanding the relationship between technology and politics allows for longer-term strategies as well as short-term tactics in building cities that support good public health, and human and ecological flourishing.

How cities relate to nature is a political, ethical and technical choice. Infrastructures reflect shared values and priorities in achieving sustainable development. Is water a natural resource to be refined and distributed to meet endlessly expanding demand? Is it a scarce resource best allocated using the market? Is it a threat to human settlements and development, a risk to be managed? Is it a habitat shared by other species and the basis of healthy ecosystems? Is it the source of spiritual healing and reflection? Is it a human right, essential for good public health? How technologies are developed and deployed now and in the future depends upon the questions asked and the stories told about water in cities, as much as on the technical calculations and physical properties of science and engineering. Understanding and achieving urban water sustainability requires the capacity to discuss technologies, values and nature together.

Frameworks

Different infrastructure and technology choices have different costs and benefits to different people and the environment. Different theoretical and political frameworks understand those relationships in different ways. To make better decisions, design better infrastructure and create useful knowledge about urban water systems, it is helpful to be able to recognise these different positions and their implications. Theories about technology and society can provide structure to analyse different developments towards and away from sustainability in urban water systems.

Understanding how theory and politics frame debates and decisions about urban water sustainability highlights diversity and fragmentation within a relatively recently established field of research and practice. Identifying alternate framings of sustainability and technology may help to explain breakdowns in interdisciplinary and cross-sectoral research and practice. Disciplinary and sectoral silos have long been identified as obstacles to integrated approaches. This is commonly talked about as a language barrier, with each discipline having its own exclusive, specialist terminology. Jargon undoubtedly makes communication difficult, but interdisciplinary ventures still falter even when care is taken to speak in plain language. Professional and academic disciplines not only have their own languages; they also have their own frameworks – shared meanings and stories about how their knowledge contributes to improving the world. Misunderstanding and conflict can arise when frameworks are misaligned. Identifying the most common discursive and conceptual frameworks that underpin alternative narratives may help to explain, if not resolve, some of the challenges of interdisciplinary work in urban sustainability.

Urban water sustainability is simultaneously a unifying proposition for a progressive, positive future and a set of divergent strategies for social, political and technical transformation. As such, it reflects wider debates within environmental politics. Five distinct but overlapping frameworks can be identified in urban water sustainability – sustainable development, ecological modernisation, socio-technical systems, urban political ecology and radical ecology. Sustainable development is the familiar framing of the need to deliver the benefits of development to the global population within ecological and resource limits (WCED, 1987). Ecological modernisation promises that environmental problems can be solved by reforming the institutions of modern society, with a central role for technological innovation (Mol, 2000). Socio-technical systems emphasise the co-evolution of technology and society in identifying opportunities to transition towards more sustainable infrastructure (Geels, 2002; Vliet et al., 2005). Political ecology frames sustainability as a socio-environmental problem and highlights the connections between social inequality and ecological degradation (Swyngedouw et al., 2002). Radical ecology makes the case for fundamentally reconfiguring relationships between nature and society, emphasising the value of nature in its own right (Merchant, 1992). Frameworks can be used in three ways: as an analysis tool, to identify assumptions underlying various propositions for sustainability; as a normative standpoint, to define how sustainability should be achieved according to a particular set of values; and pragmatically, to align propositions for change within a dominant political and theoretical discourse.

Technologies

The role of technology and infrastructure in sustainability is contested. For some critics, modern technology is depicted as the source of pollution, as responsible for the exploitation of people and natural resources and for the destruction of ecosystems (Foreman and Haywood, 1993; Meadows et al., 1972). For technological optimists, environmental degradation and economic inequalities are side effects of incomplete modernisation, and technology is the means to improve resource efficiency and reduce pollution (Asafu-Adjaye et al., 2015). Most people recognise that technology is both a cause and a potential solution to environmental problems, and that it is necessary but not sufficient to achieve sustainability. There remains significant diversity in how the role of technology and infrastructure are understood in different sustainability frameworks. Characterising these different perspectives helps one to understand broader debates, policies and approaches to urban water sustainability.

The complexity of urban water sustainability may be revealed by analysing the main trends and technologies that have emerged in water infrastructure since the 1960s. This book addresses five technical trends and drivers in future urban water management – demand, sanitation, drainage, reuse and desalination. Focussing on these five specific categories of technology and water management grounds the analysis in a set of identifiable, tangible, material changes that are underway in cities around the world.

Classical approaches to urban infrastructure are founded on expanding provision to meet growing demand (Butler and Memon, 2005). Demand for water and sanitation services remains unmet in many cities in the Global South, with disastrous consequences for health and economic development (UNICEF and WHO, 2015). Beyond meeting basic needs, infrastructure has enabled the development of lifestyles and cultures that continue to drive up demand for resources. The challenge of sustainability is to ensure that demand is met within resource constraints. Understanding the drivers of demand and opportunities to reduce per capita water use in cities has become a major focus of urban water management. Approaches to managing demand include implementing more efficient water-using technologies, economic incentives to reduce demand through metering and pricing, behaviour change programmes based on understanding individual motivations and attitudes, and wider cultural change to better align everyday life with environmental contexts (Hoolohan and Browne, 2016).

Sanitation infrastructure is thought to have delivered the greatest improvements in public health of any urban development in the last 200 years (Ferriman, 2007). Water-based sanitation has transformed cities in the Global North, eliminating risk of diseases such as cholera, typhoid and dysentery (Melosi, 2008). Flushing toilets account for around a third of urban water demand, and wastewater discharge to the environment is a significant source of pollution. Constructing sewer networks and wastewater treatment plants also requires large capital investment, and delivering sewerage networks to slums and informal settlements has proven difficult. Waterless sanitation has emerged as a technical alternative, promising resource recovery and lower-cost solutions to public health and environmental crises (Langergraber and Muellegger, 2005). However, significant challenges remain in delivering this alternative model of sanitation at scale, including sludge management, risk management and business models.

Surface water drains are one of the oldest urban infrastructures. Conventional approaches to drainage are based on fast conveyance of surface water, to remove rainwater from the urban environment as quickly as possible. Stormwater drains and overflows from sewers that combine surface and wastewater are a major source of urban pollution. Since the 1970s more sustainable approaches have emerged that aim to emulate natural hydrology, slowing the flow of water, providing space for water in cities through integrated green infrastructure (Wong and Brown, 2009). In addition to reducing pollution and providing flood protection, sustainable drainage claims to deliver additional benefits to cities, including cooling, biodiversity and generally better environmental quality (Grant, 2016).

Reuse and recycling are common tenets of sustainability, and water is the ultimate recyclable material. Water reuse takes many forms – potable and non-potable, direct and indirect, centralised and decentralised, planned and unplanned (Wilcox et al., 2016). Sustainable water management promotes the concept of 'fit-for-purpose water' in contrast to the standard practice of using highly treated drinking water for all uses. This enables reuse of greywater, stormwater and rainwater without energy-intensive treatment, but presents a new set of risks. Reused urban wastewater is an important source of water for irrigation in peri-urban and rural agricultural areas (Miller-Robbie et al., 2017). Wastewater may also be recycled into drinking water supply, requiring high standards of treatment through energy-intensive technologies (Guo and Englehardt, 2015). Potable reuse of wastewater has been the subject of significant public controversy (Marks, 2006). Reuse is not a singular, sustainable technological solution to urban water shortages, but a complex set of strategies with varying social, environmental and economic consequences.

Desalination is commonly presented as the ultimate technical solution to water shortages. Creating freshwater from salt water would seem to open up a limitless new supply of water to meet endlessly expanding demand. Desalination is costly and energy intensive, and it has localised impacts on aquatic environments. It is highly controversial within urban water sustainability. Proponents claim that desalination powered by low-carbon energy sources is the ultimate sustainable technology (Ghaffour et al., 2015). Critics maintain that desalination undermines broader efforts towards sustainability, including demand management and reuse, and as a high-cost, capital-intensive infrastructure option it is driven as much by the needs of global investment and technology firms as it is by water scarcity (March, 2015).

Integration of different infrastructures and sectors has been a key theme in urban water management in recent decades. This book aims to improve understanding of the different elements of water infrastructure, rather than emphasising the need for integration into a comprehensive urban water system. Inevitably points of intersection become evident across the different technologies and trends, but that is not the purpose of this analysis. Working through technologies provides a bottom-up view of the elements of urban water infrastructure that contribute to and undermine sustainability, and provides a physical grounding for political and technical discourse.

Looking at technologies through particular frameworks provides a viewpoint from which to see different social and political values and debates about urban water sustainability. Technologies are important because they open up new possibilities for water and sanitation in cities. Technology is also important because it can show the core values shaping cities. Analysing water technology and infrastructure reveals much about relationships between cities, people and nature.

Reading this book

This book analyses technical developments in water infrastructure using five frameworks of sustainability. Chapter 2 reviews the concept of urban water sustainability as it developed through global discourse to agreed principles for integrated water management and water sensitive cities. Chapter 3 highlights the complexity of infrastructure in general, as the intersection of social, political, economic, ecological and technical elements of cities. It addresses the invisibility of infrastructure as background to everyday life, urban life, trends in ownership and investment, decentralisation and the significance of technologies and urban planning in gendered constructions of the city. The five frameworks for analysing water sustainability are outlined in Chapter 4: sustainable development, ecological modernisation, sociotechnical systems, political ecology and radical ecology.

Chapters 5–9 review key developments in water demand, sanitation, drainage, reuse and desalination. Each of these technical options is outlined in sufficient detail to give an understanding of its historical context and function in cities and modern development. These chapters might be read in isolation as technical primers on recent trends in urban water infrastructure. Each technical trend is analysed using the five frameworks to draw out different perspectives on its role in achieving urban water sustainability. This reveals consensus, complementarity, contradiction and critique across different political and theoretical positions. The conclusion draws key insights from this analysis for the broader goal of sustainable cities.

The intention of the book is to provide structure to discussions and debates about the role of technology in achieving urban water sustainability. The goal is to provide the means for discussing technical, social and ecological factors that constitute water infrastructure, across different disciplinary, professional and political boundaries. It neither provides a blueprint for sustainable cities or water systems nor a road map for implementation. It aims to enable clearer analysis of options for the future of cities and water, within academic research, professional practice and democratic debate and decision-making.

References

Allan, J.A. 2005. Water in the Environment/Socio-Economic Development Discourse: Sustainability, Changing Management Paradigms and Policy Responses in a Global System. *Government and Opposition* 40, 181–199. doi:10.1111/j.1477–7053.2005.00149.x

Allan, J.A. 2006. IWRM: The New Sanctioned Discourse? in: Mollinga, P.P., Dixit, A. and Athukorala, K. (Eds.), *Integrated Water Resources Management*. Sage, New Delhi and London, pp. 38–63.

Allan, T. 2011. *Virtual Water: Tackling the Threat to Our Planet's Most Precious Resource*. I.B.Tauris, London and New York.

Arnell, N.W. 2004. Climate Change and Global Water Resources: SRES Emissions and Socio-Economic Scenarios. *Global Environmental Change, Climate Change* 14, 31–52. doi:10.1016/j.gloenvcha.2003.10.006

Asafu-Adjaye, J., Blomqvist, L., Brand, S., Brook, B., Defries, R., Ellis, E., Foreman, C., Keith, D., Lewis, M., Mark, L., Nordhaus, T., Pielke Jr, R., Pritzker, R., Roy, J., Sagoff, M., Shellenberger, M., Stone, R. and Teague, P. 2015. *An Ecomodernist Manifesto*.

Bakker, K. 2010. *Privatizing Water*. Cornell University Press, Ithaca and London.

Brown, R., Keath, N. and Wong, T. 2009. Urban Water Management in Cities: Historical, Current and Future Regimes. *Water Science and Technology* 59(5), 847–855.

Butler, D. and Memon, F.A. (Eds.) 2005. *Water Demand Management*. IWA Publishing, London.

Cook, C. and Bakker, K. 2012. Water Security: Debating an Emerging Paradigm. *Global Environmental Change* 22, 94–102. doi:10.1016/j.gloenvcha.2011.10.011

Ferriman, A. 2007. BMJ Readers Choose the "Sanitary Revolution" as Greatest Medical Advance Since 1840. *BMJ: British Medical Journal* 334, 111.

Foreman, D. and Haywood, B. 1993. *Ecodefense: A Field Guide to Monkeywrenching*. Abbzug Press, Chico, CA.

Geels, F.W. 2002. Technological Transitions as Evolutionary Reconfiguration Processes: A Multi-Level Perspective and a Case-Study. *Research Policy, NELSON + WINTER + 20* 31, 1257–1274. doi:10.1016/S0048-7333(02)00062–00068

Ghaffour, N., Bundschuh, J., Mahmoudi, H. and Goosen, M.F.A. 2015. Renewable Energy-Driven Desalination Technologies: A Comprehensive Review on Challenges and Potential Applications of Integrated Systems. *Desalination, State-of-the-Art Reviews in Desalination* 356, 94–114. doi:10.1016/j.desal.2014.10.024

Grant, G. 2016. *The Water Sensitive City*. Wiley-Blackwell, Chichester and West Sussex.

Guo, T. and Englehardt, J.D. 2015. Principles for Scaling of Distributed Direct Potable Water Reuse Systems: A Modeling Study. *Water Research* 75, 146–163. doi:10.1016/j.watres.2015.02.033

Hoolohan, C. and Browne, A.L. 2016. Reframing Water Efficiency: Determining Collective Approaches to Change Water Use in the Home. *British Journal of Environment and Climate Change* 6, 179–191.

International Water Association. 2015. *The IWA Principles for Water-Wise Cities* [WWW Document]. International Water Association. www.iwa-network.org/projects/water-wise-cities/.

Langergraber, G. and Muellegger, E. 2005. Ecological Sanitation – A Way to Solve Global Sanitation Problems? *Environment International* 31, 433–444. doi:10.1016/j.envint.2004.08.006

March, H. 2015. The Politics, Geography, and Economics of Desalination: A Critical Review. *WIREs Water* 2, 231–243. doi:10.1002/wat2.1073

Marks, J.S. 2006. Taking the Public Seriously: The Case of Potable and Non Potable Reuse. *Desalination* 187, 137–147. doi:10.1016/j.desal.2005.04.074

Marlow, D.R., Moglia, M., Cook, S. and Beale, D.J. 2013. Towards Sustainable Urban Water Management: A Critical Reassessment. *Water Research, Urban Water Management to Increase Sustainability of Cities* 47, 7150–7161. doi:10.1016/j.watres.2013.07.046

Meadows, D.H., Meadows, D.L., Randers, J. and Behrens, W.W. 1972. *Limits to Growth*. Universe Books, New York.

Melosi, M.V. 2008. *The Sanitary City: Environmental Services in Urban America from Colonial Times to the Present*. University of Pittsburgh Press.

Merchant, C. 1992. *Radical Ecology: The Search for a Livable World*. Routledge, London and New York.

Miller-Robbie, L., Ramaswami, A. and Amerasinghe, P. 2017. Wastewater Treatment and Reuse in Urban Agriculture: Exploring the Food, Energy, Water, and Health Nexus in Hyderabad, India. *Environmental Research Letters* 12, 75005. doi:10.1088/1748–9326/aa6bfe

Mitchell, V. 2006. Applying Integrated Urban Water Management Concepts: A Review of Australian Experience. *Environmental Management* 37, 589–605. doi:10.1007/s00267-004-0252-1

Mol, A. 2000. *Ecological Modernisation Around the World: Perspectives and Critical Debates*. London: Taylor & Francis.

Niemczynowicz, J. 1999. Urban Hydrology and Water Management – Present and Future Challenges. *Urban Water* 1, 1–14. doi:10.1016/S1462–0758(99) 00009–00006

Novotny, V., Ahern, J. and Brown, P. 2010. *Water Centric Sustainable Communities*. John Wiley and Sons, Hoboken.

Oki, T. and Kanae, S. 2006. Global Hydrological Cycles and World Water Resources. *Science* 313, 1068–1072. doi:10.1126/science.1128845

Russo, T., Alfredo, K. and Fisher, J. 2014. Sustainable Water Management in Urban, Agricultural, and Natural Systems. *Water* 6, 3934–3956. doi:10.3390/w6123934

Shiklomanov, I.A. 2000. Appraisal and Assessment of World Water Resources. *Water International* 25, 11–32. doi:10.1080/02508060008686794

Shove, E. 2004. *Comfort, Cleanliness and Convenience: The Social Organization of Normality*. Berg Publishers, Oxford.

Swyngedouw, E., Kaika, M. and Castro, E. 2002. Urban Water: A Political-Ecology Perspective. *Built Environment (1978)* 28, 124–137.

UN-HABITAT. 2003. *Water and Sanitation in the World's Cities*. Earthscan, London and Sterling, VA.

UNICEF and WHO. 2015. *Progress on Sanitation and Drinking Water – 2015 Update and MDG Assessment*. WHO Press, Geneva.

United Nations Human Settlements Programme. 2006. *The State of the World's Cities 2006/2007 : The Millennium Development Goals and Urban Sustainability : 30 Years of Shaping the Habitat Agenda*. Earthscan London and Sterling, VA.

Vliet, B.V., Shove, E. and Chappells, H. 2005. *Infrastructures of Consumption: Environmental Innovation in the Utility Industries*. Earthscan, London and Sterling, VA.

WCED. 1987. *Towards Sustainable Development, in: Our Common Future*. Oxford University Press, Oxford and New York, pp. 43–65.

Wilcox, J., Nasiri, F., Bell, S. and Rahaman, M.S. 2016. Urban Water Reuse: A Triple Bottom Line Assessment Framework and Review. *Sustainable Cities and Society* 27, 448–456. doi:10.1016/j.scs.2016.06.021

Wong, T.H.F. and Brown, R.R. 2009. The Water Sensitive City: Principles for Practice. *Water Science and Technology* 60, 673–682. doi:10.2166/wst.2009.436

Zeitoun, M., Lankford, B., Krueger, T., Forsyth, T., Carter, R., Hoekstra, A.Y., Taylor, R., Varis, O., Cleaver, F., Boelens, R., Swatuk, L., Tickner, D., Scott, C.A., Mirumachi, N. and Matthews, N. 2016. Reductionist and Integrative Research Approaches to Complex Water Security Policy Challenges. *Global Environmental Change* 39, 143–154. doi:10.1016/j.gloenvcha.2016.04.010

2 Water and sustainable cities

Introduction

Water is a vital element of any city, and a key challenge for sustainability. Secure access to water and sanitation are essential for human development (UN-HABITAT, 2003). Although mostly renewable, freshwater resources are limited and water pollution is a serious environmental and public health problem (United Nations Economic and Social Council, 1997). Abstracting water from rivers and aquifers and discharging surface and wastewater into the environment are some of the most significant impacts that cities have on their surrounding environments. Water is also an important element of urban environments, providing opportunities for recreation, habitat for biodiversity, and general amenity. Water is a source of risk from flooding, pollution and waterborne disease.

As an issue of resource management and pollution control, water holds much in common with other urban infrastructure sustainability issues. In the Global North water is typically delivered through complex distribution networks, requiring high levels of technical expertise and strong governance. Ensuring efficient use of the resource and the infrastructure network that delivers it holds some similarities with energy and transport. Managing water pollution may also be thought of as part of environment policy and strategy, in common with waste and air pollution.

Water has particular physical, ecological and biological characteristics that set it apart from other infrastructure issues. It is critical for basic health and wellbeing. Water persists in cities through rainfall, runoff, wetlands, ponds and streams as a viable agent in landscapes and places. Water is heavy, difficult to transport over long distances. Water has shaped cities and cultures since ancient times.

Coming to terms with both the general and specific sustainability challenges for urban water has been a theme in global discourse, national and local policy, academic research and professional practice. This chapter describes the rise of water and cities in global debates about sustainable development, the principles of integrated and sustainable urban water management, and the concept of water sensitive cities.

International water deliberation

Water has been a feature of United Nations conferences and conventions since the 1970s, with changing emphasis and priorities through the decades. Following the landmark 1972 UN Conference on the Human Environment in Stockholm, the 1977 UN Conference on Water in Mar del Plata, Argentina, showed the importance of water to the environmental movement and human development but made little reference to cities. The conference resolution addressed assessment of water resources and community water supply. Sanitation was addressed as an issue aligned with water supply and public health, but there was no explicit recognition of integrated water management. This approach underpinned the 1981–1990 'International Drinking Water Supply and Sanitation Decade', which emphasised the importance of water provision for public health and development. The 1987 UN Commission for Environment and Development Report, *Our Common Future*, famously defined the concept of sustainable development as 'development that meets the needs of the present without compromising the ability of future generations to meet their own needs', but it did not address water as a specific area of focus (WCED, 1987). Instead it identified the importance of water and sanitation provision to urban development and the need to manage water resources to meet growing demands for food.

Integration of water management and the importance of water in sustainable urbanisation were explicitly recognised in international statements and agreements by the early 1990s. In January 1992 the International Conference on Water and the Environment in Dublin outlined four guiding principles for water and sustainable development:

1   Freshwater is a finite and vulnerable resource, essential to sustain life, development and the environment
2   Water development and management should be based on a participatory approach, involving users, planners and policy-makers at all levels
3   Women play a central part in the provision, management and safeguarding of water
4   Water has an economic value in all its competing uses and should be recognised as an economic good

Sustainable urban development was one of 11 specific areas of action in the Dublin Statement, which informed discussion about water at the UN Conference on Environment and Development, or 'Earth Summit', held in Rio de Janeiro later that year. Chapter 18 of the UNCED resolution, known as Agenda 21, addressed 'Protection of the Quality and Supply of Freshwater Resources: Application of Integrated Approaches to the Development, Management and Use of Freshwater Resources' (United Nations Sustainable Development, 1992, p. 21). Agenda 21 set the basis for integration of water management, and it included a programme for 'Water and sustainable urban development'. It outlined an approach that integrates supply, demand, sanitation, drainage and flood protection, and that is based on full public participation in water management policy and decision-making. Water is discussed elsewhere in Agenda 21, in relation to improving urban health and integrated provision of environmental infrastructure for sustainable human settlement development.

The 1997 'Comprehensive assessment of the freshwater resource of the world' by the UN Commission on Sustainable Development provided a forecast of increasing water scarcity, with the often quoted estimate that 'by the year 2025 as much as two thirds of the world population could be living under stress conditions' (United Nations Economic and Social Council, 1997). The report found that water scarcity and pollution causes widespread public health problems, slows economic and agricultural development and damages ecosystems. Urban development was included as a contributing factor in water scarcity and pollution. The report addressed the need to manage demand for water, including improving water efficiency, and to reduce water pollution, particularly from urban wastewater. It highlighted the importance of public participation in driving improved water management, particularly the role of women.

Gender has been a consistent theme in international negotiations and sustainable development programmes to improve access and management of water and sanitation. Around the world, lack of access to water and sanitation disproportionately affects women. Women are usually responsible for collecting water in communities without piped supply and for caring for those affected by water-related diseases (Ahlers and Zwarteveen, 2009). Poor access to sanitation can leave women vulnerable to violence and harassment, and lack of provision of toilets for girls can result in low attendance at school (Abrahams et al., 2006). Given the importance of water and sanitation provision for women, it has been widely recognised that women should participate directly in decision-making about infrastructure and water resources, and women and gender are recognised as specific stakeholders and issues to be addressed in all major international statements about water from the 1990s forward.

In 2000 water and sanitation were included in the UN Millennium Declaration and Millennium Development Goals (MDGs). The Declaration addressed access to safe drinking water as an issue of poverty alleviation and recognised the need for better management to stop unsustainable exploitation of water resources (United Nations General Assembly, 2000). Water and sanitation were included within Goal 7 of the MDGs, 'Ensure environmental sustainability', with Target 10 being 'Halve, by 2015, the proportion of people without sustainable access to safe drinking water and sanitation'. Redoubling efforts in the 1980s, 2005–2015 was the 'International Decade for Action: Water for Life'. Again, activities emphasised water and sanitation for public health and poverty alleviation, including the need for sustainable management of resources and water quality. According to official UN data the target for access to water was met before the 2015 deadline, but very little progress was made in improving the proportion of people with access to sanitation.

The 2002 UN Summit on Sustainable Development in Johannesburg built on previous approaches, producing a Framework for Action on Water and Sanitation. The nine action areas of the framework covered the MDG targets for water and sanitation, IWRM, improved productivity of agriculture, safeguarding human health, improving disaster preparedness, strengthening institutional and technical capacity, investment and protecting water quality.

The impacts of climate change on water resources and flooding have been a major concern in national and international science and policy since the 1990s. Some of the most significant impacts of climate change are likely to be felt in water resources management and flooding. Changing patterns of precipitation will alter the availability of water for human use. In particular, reduced overall average precipitation could contribute to physical water scarcity in some regions, and greater variability could lead to more frequent, prolonged and intense droughts and floods (Bates et al., 2008; IPCC, 2014). Climate change will impact water availability in different ways in different regions, with significant impacts forecast in glacier-fed river systems, but globally increases in water demand are likely to contribute to water scarcity to a greater extent than changing precipitation due to climate change (Arnell, 2004).

In 2010 the UN General Assembly formally recognised the Human Right to Water and Sanitation (United Nations General Assembly, 2010). Access to these services is no longer simply a beneficial outcome of development or necessary for economic growth and public health, but is a basic human right that is necessary for the realisation of all other human rights. The declaration provides added impetus for campaigns and policy, particularly given the lack of progress in delivering the MDG for sanitation. It calls on nations and international organisations to provide necessary investment and governance to deliver these services required to meet this right. Whilst progress may be limited in some countries, the Human Right to Water and Sanitation confirmed the importance of these issues to basic human dignity and their importance in progress towards sustainable development.

The Sustainable Development Goals, which replaced the MDGs in 2015, include Goal 6, 'Ensure availability and sustainable management for water and sanitation for all', as one of 17 headline goals (United Nations, 2015). The SDG for water includes eight targets relating to access, integrated management, pollution, efficiency, ecosystem restoration, participation and governance. Reducing the impact of water-related disasters is also included in SDG 11, relating to safe, sustainable and resilient cities.

The 2015 United Nations World Water Development Report, *Water for a Sustainable World*, addressed the water SDGs and the main elements of water and sustainable development (United Nations World Water Assessment Programme, 2015). With forecasts for increasing demand for water from cities and the sustainable development challenges of urban population growth, the report included a specific chapter on urbanisation. It highlighted the challenge of water and sanitation provision to slums, particularly in contexts where local authorities are reluctant to recognise informal settlements through the provision of services. Other challenges include governance, climate change, pollution, and wastewater treatment. Responses aligned with the SDGs were proposed to include pro-poor water and sanitation policies, integrated urban water management, sustainable sanitation, improvement of governance and adaptation to climate change and water-related disasters.

Evolving as part of the international discourse of sustainable development, the concept of urban water sustainability can then be said to be founded on the following principles:

- access to water and sanitation is a human right and is essential for development and good public health (United Nations General Assembly, 2010);
- freshwater is a limited resource which should be protected from over-exploitation and pollution (United Nations Economic and Social Council, 1997);
- the public, particularly women, should participate in planning and decisions about water (International Conference on Water and the Environment, 1992); and
- water resources management should integrate different sectors and systems, including agriculture, industry and domestic demand, drinking water, sanitation and surface water and drainage (Marseille Statement, 2001).

Integrated water management

Integration has been a central theme of policies for sustainable development of water since the 1990s, though the idea of integration can be traced back to the middle of the twentieth century (Biswas, 2004). Integrated water resources management (IWRM) aims to manage competing uses of water at catchment scale. The Global Water Partnership (2000) defined IWRM as a

> process which promotes the co-ordinated development and management of water, land and related resources, in order to maximize the resultant economic and social welfare in an equitable manner without compromising the sustainability of vital ecosystems.
>
> (Agarwal et al., 2000)

IWRM is intended to provide for sustainable management of limited, local water resources, allocating water to different agricultural users, industry and municipal supply, and managing land and water together to improve ecological sustainability and resilience. IWRM policies aim to involve stakeholders in decision-making and include consideration of using water infrastructure to manage flooding as well as water supply, particularly using dams and river basin transfers to manage flows during flood events. Urban utilities are typically powerful stakeholders in catchments, with allocations for public health requirements for cities determined to be critical compared with irrigation for agricultural production.

Integrated urban water management (IUWM) translates this approach to the city scale. The emphasis is on managing water supply, wastewater treatment, drainage and flood control to achieve mutual benefits and sustainable development (Bahkri, 2012; Brown et al., 2009; Mitchell, 2006). Following the principles of stakeholder engagement and public participation, IUWM involved different sections of water and wastewater utilities, local government, planners, developers, industry and civil society, bridging urbanisation processes and water management. IUWM aims to improve the efficiency of water use, identify opportunities to reduce reliance on external resources, and minimise the volume of water to be transferred and treated in wastewater and stormwater systems. IUWM literature speaks of the need to understand cities as water catchments and to move from single-use, linear flows to circular patterns which reuse water within the city rather than disposing of wastewater and surface water into the environment (Andrew, 2007; Bahkri, 2012).

Integrated management of water also requires consideration of the intersection between water and other systems and services. The connections between water for food and energy production have been embedded in the principles of integrated management since the 1980s (Biswas, 2004). The relationship between water, energy and food systems has been characterised as a nexus, recognising the energy requirements of water infrastructure in addition to the water requirements for energy and food (Dodds and Bartram, 2016; Miller-Robbie et al., 2017). At the catchment-scale, demand for water for agricultural irrigation to produce food is a key component of catchment-based integrated management, which also requires energy for pumping. Hydropower and water for cooling of thermal power stations may be a large local source of industrial water demand. On the other side of the nexus, energy is required for water and wastewater treatment and distribution.

Energy intensity is particularly important in assessing the sustainability of alternative water supply and treatment technologies. Smaller-scale distributed water and wastewater treatment technologies may require lower energy for distribution, but may be less energy efficient than centralised treatment plants (Sapkota et al., 2014). Membrane-based treatment technologies for desalination and water reuse are significantly more energy intensive than conventional treatment technologies in most cases (Cooley and Wilkinson, 2012). These technologies may help to overcome water scarcity, but increasing demand for energy can undermine the overall sustainability of the system. The energy requirements of wastewater treatment are also important in urban water sustainability (Balkema et al., 2002). Unreliable local electricity supply in cities in the Global South can be a significant factor undermining the effectiveness of conventional wastewater treatment plants. The energy requirements of treatment technologies for removal of micro-pollutants, such as pesticides and pharmaceuticals, can lead to significant increase in energy demand for water systems in the Global North (Joss et al., 2008).

The concept of 'virtual water' or 'water footprint' provides an indicator of the volume of water required to produce food and industrial products. Developed by Tony Allen and others, virtual or 'embodied' water analysis shows that consumption of food and other materials is associated with much higher volumes of water than the volume of water consumed in everyday activities such as showering, drinking and toilet flushing (Allan, 2011; Chapagain and Hoekstra, 2004). An espresso coffee requires 140 litres of water to produce, including water to grow and process the coffee beans, which is close to the average domestic consumption of water in the UK of 150 litres per person per day (Allan, 2005; DEFRA, 2011). Virtual water also provides the possibility of overcoming local water scarcity through international trade (Hanasaki et al., 2010). Cities in water-scarce regions are able to import food produced in other parts of the world. Unfortunately, agricultural and industrial production for export can be the cause of water scarcity and unsustainable exploitation of water resources, and the water footprint of products helps to reveal wider impacts of consumption.

The goal of integration in water management has been criticised as vague, unachievable in practice and vulnerable to powerful interests at the cost of small landowners and people living in poverty (Allan, 2006; Biswas, 2004; Mollinga et al., 2006). IWRM and IUWM are subject to similar criticisms as sustainable development, resilience and other concepts that are difficult to precisely define and measure. Stakeholder participation processes and attention to the economic value of water may reinforce the power of large corporate actors in the catchment, to the detriment of smaller landholders and water users. The concept of integration itself inherently increases the complexity of urban water systems and their connection to other infrastructures and urban processes. This presents substantive challenges to engineering and management of urban water systems, as integrating theories, data and modelling is not trivial (Bach et al., 2014; Makropoulos et al., 2008). Integration has emerged as the preferred approach to understanding urban water infrastructure and resources but remains a difficult goal to implement (Brown and Farrelly, 2009).

Water sensitive cities

The concept of the 'water sensitive city' bridges developments in water and urban sustainability. According to Wong and Brown (2009), the three pillars of the water sensitive city are:

Cities as water supply catchments
Cities providing ecosystem services
Cities comprising water sensitive communities

Conventional urban water infrastructure has separated people from natural systems and hidden the flows of water through the city and buildings. Water sensitive cities aim to reconnect the urban and natural environments through local hydrology and communities. This approach is consistent with the principles for landscape design developed by Ian McHarg (1995) in the 1960s in *Design with Nature*. McHarg's approach to urban design involved building up layers of spatial information, beginning with the local geology and soil types, followed by local hydrology, ecology and other elements to design places that were consistent with and enhanced the local environment. Water is a fundamental element of natural environments and in water sensitive cities it becomes a unifying element of urban environments.

The movement for water sensitive cities and communities extends earlier efforts in water sensitive urban design, low-impact development and sustainable drainage systems that focussed on sustainable approaches to stormwater management. Since the 1970s engineers and urban designers have developed principles and techniques that mimic natural catchment hydrology in how rainwater that falls on cities is stored, flows, infiltrates the ground and is passively treated before reaching local rivers and other receiving waters (see Chapter 7).

In the water sensitive city these ideas have been extended to include opportunities for utilising water captured, stored and recycled within the city as an alternative supply, reducing pressure on conventional water resources (Grant, 2016; Novotny et al., 2010). Rainwater harvested from roofs can be used for non-potable uses such as toilet flushing and landscape irrigation, with minimal treatment, and can also be treated to potable standards (Campisano et al., 2017). Stormwater harvesting is also a source of water for non-potable use, abstracting and storing water from drains in cities where sewage and stormwater are transferred and treated separately. The water sensitive city also considers opportunities for greywater reuse and provides for dual supply of potable and non-potable water at building and neighbourhood scale (Ferguson et al., 2013).

Green infrastructure, such as green roofs, parks, rain gardens, wetlands and ponds, provides opportunities to manage water by providing local sites for infiltration, for storage and to attenuate flows, and also provides a range of other benefits (Dover, 2015). Green infrastructure can provide habitats for local wildlife and enrich plant diversity. Through evapotranspiration and shading it can contribute to the cooling of cities, alleviating the urban heat island effect. Green infrastructure has been demonstrated to improve mental and physical health and wellbeing, generally increasing the quality of the urban environment. Food production can also be incorporated into green infrastructure systems, providing additional benefits.

Where green infrastructure and sustainable water systems are integrated into good urban design and planning, they can form the basis of water sensitive communities (Dean et al., 2016; Ferguson et al., 2013). Engaging communities in the planning and design of water and green infrastructure draws on and enhances local knowledge about water and the urban environment. This requires water to be considered from the outset of planning and design for urban development and regeneration, allowing for neighbourhood and building design that accounts for local hydrology, topology and green infrastructure opportunities, as well as the needs and aspirations of residents (Lerer et al., 2015).

Water and sustainable cities

Water has been included as an element in sustainable city designs and debates, though often with a lower profile and at a later stage in the process than energy, transport, pollution and waste. It has been an important topic for international deliberation and goals for sustainable development. At a global level, cities impact water resources more through consumption of food and industrial products, but at a local level urban water infrastructure has significant impacts on the environment. Water plays an important role in the health and prosperity of cities. It provides wider connections to urban life, design and politics, beyond improving the efficiency and integration of water and sanitation infrastructure.

Sustainable urban water systems support sustainable cities. Water infrastructure shapes the way that people in cities relate to the environment. As such, urban water sustainability is not simply about averting water resource crises or redressing the unsustainability of conventional infrastructure, but realising the opportunities for more harmonious, resilient relationships between people and nature within and beyond cities.

References

Abrahams, N., Mathews, S. and Ramela, P. 2006. Intersections of "Sanitation, Sexual Coercion and Girls' Safety in Schools". *Tropical Medicine & International Health* 11, 751–756. doi:10.1111/j.1365–3156.2006.01600.x

Agarwal, A., delos Angeles, M., Bhatia, R., Cheret, I., Davila-Poblete, S., Falkenmark, M., Jonch-Clausen, T., Ait Khadi, M., Kindler, J., Rees, J., Roberts, P., Rogers, P., Solanes, M. and Wright, A. 2000. *Integrated Water Resources Management*. Global Water Partnership, Stockholm.

Ahlers, R. and Zwarteveen, M. 2009. The Water Question in Feminism: Water Control and Gender Inequities in a Neo-Liberal Era. *Gender, Place & Culture* 16, 409–426. doi:10.1080/09663690903003926

Allan, J.A. 2005. Water in the Environment/Socio-Economic Development Discourse: Sustainability, Changing Management Paradigms and Policy Responses in a Global System. *Government and Opposition* 40, 181–199. doi:10.1111/j.1477–7053.2005.00149.x

Allan, J.A. 2006. IWRM: The New Sanctioned Discourse? in: Mollinga, P.P., Dixit, A. and Athukorala, K. (Eds.), *Integrated Water Resources Management*. Sage, New Delhi and London, pp. 38–63.

Allan, T. 2011. *Virtual Water: Tackling the Threat to Our Planet's Most Precious Resource.* I.B.Tauris, London and New York.

Andrew, S. 2007. Water and Cities – Overcoming Inertia and Achieving a Sustainable Future, in: Novotny, V. and Brown, P. (Eds.), *Cities of the Future: Towards Integrated Sustainable Water and Landscape Management : Proceedings of an International Workshop Held July 12–14, 2006 in Wingspread Conference Center.* IWA Publishing, Racine, WI, pp. 18–31.

Arnell, N.W. 2004. Climate Change and Global Water Resources: SRES Emissions and Socio-Economic Scenarios. *Global Environmental Change, Climate Change* 14, 31–52. doi:10.1016/j.gloenvcha.2003.10.006

Bach, P.M., Rauch, W., Mikkelsen, P.S., McCarthy, D.T. and Deletic, A. 2014. A Critical Review of Integrated Urban Water Modelling – Urban Drainage and Beyond. *Environmental Modelling & Software* 54, 88–107. doi:10.1016/j.envsoft.2013.12.018

Bahkri, A. 2012. *Integrated Urban Water Management.* Global Water Partnership, Stockholm.

Bakker, K. 2010. *Privatizing Water.* Cornell University Press, Ithaca and London.

Balkema, A.J., Preisig, H.A., Otterpohl, R. and Lambert, F.J.D. 2002. Indicators for the Sustainability Assessment of Wastewater Treatment Systems. *Urban Water* 4, 153–161. doi:10.1016/S1462–0758(02)00014–00016

Bates, B., Kundzewicz, Z., Palutikof, J. and Wu., S. 2008. *Climate Change and Water, Technical Paper of the Intergovernmental Panel on Climate Change.* IPCC Secretariat, Geneva.

Biswas, A.K. 2004. Integrated Water Resources Management: A Reassessment. *Water International* 29, 248–256. doi:10.1080/02508060408691775

Brown, R.R. and Farrelly, M.A. 2009. Delivering Sustainable Urban Water Management: A Review of the Hurdles We Face. *Water Science and Technology* 59, 839–846. doi:10.2166/wst.2009.028

Brown, R.R., Keath, N. and Wong, T.H.F. 2009. Urban Water Management in Cities: Historical, Current and Future Regimes. *Water Science & Technology* 59, 847. doi:10.2166/wst.2009.029

Campisano, A., Butler, D., Ward, S., Burns, M.J., Friedler, E., DeBusk, K., Fisher-Jeffes, L.N., Ghisi, E., Rahman, A., Furumai, H. and Han, M. 2017. Urban Rainwater Harvesting Systems: Research, Implementation and Future Perspectives. *Water Research* 115, 195–209. doi:10.1016/j.watres.2017.02.056

Chapagain, A.K. and Hoekstra, A.Y. 2004. *Water Footprints of Nations Volume 1: Main Report.* UNESCO – IHE.

Cook, C. and Bakker, K. 2012. Water Security: Debating an Emerging Paradigm. *Global Environmental Change* 22, 94–102. doi:10.1016/j.gloenvcha.2011.10.011

Cooley, H. and Wilkinson, R. 2012. *Implications of Future Water Supply Sources for Energy Demands.* Water Reuse Association, Alexandria.

Dean, A.J., Lindsay, J., Fielding, K.S. and Smith, L.D.G. 2016. Fostering Water Sensitive Citizenship – Community Profiles of Engagement in Water-Related Issues. *Environmental Science & Policy* 55, Part 1, 238–247. doi:10.1016/j.envsci.2015.10.016

DEFRA. 2011. *Water for Life.* Stationery Office, London.

Dodds, F. and Bartram, J. (Eds.) 2016. *The Water, Food, Energy and Climate Nexus: Challenges and an Agenda for Action.* Routledge, London and New York.

Dover, J.W. 2015. *Green Infrastructure: Incorporating Plants and Enhancing Biodiversity in Buildings and Urban Environments*. Routledge, London and New York.

Ferguson, B.C., Frantzeskaki, N. and Brown, R.R. 2013. A Strategic Program for Transitioning to a Water Sensitive City. *Landscape and Urban Planning* 117, 32–45. doi:10.1016/j.landurbplan.2013.04.016

Grant, G. 2016. *The Water Sensitive City*. Wiley-Blackwell, Chichester and West Sussex.

Hanasaki, N., Inuzuka, T., Kanae, S. and Oki, T. 2010. An Estimation of Global Virtual Water Flow and Sources of Water Withdrawal for Major Crops and Livestock products Using a Global Hydrological Model. *Journal of Hydrology, Green-Blue Water Initiative (GBI)* 384, 232–244. doi:10.1016/j.jhydrol.2009.09.028

International Conference on Water and the Environment. 1992. *The Dublin Statement on Water and Sustainable Development*, in: International Conference on Water and the Environment. Dublin, Ireland. http://www.wmo.int/pages/prog/hwrp/documents/english/icwedece.html

IPCC. 2014. *Climate Change 2014: Impacts, Adaptation and Vulnerability. Part A: Global and Sectoral Aspects. Contribution of Working Group II to the Fifth Assessment Report of the Intergovernmental Panel on Climate Change*. Cambridge University Press, Cambridge and New York.

Joss, A., Siegrist, H. and Ternes, T.A. 2008. Are We About to Upgrade Wastewater Treatment for Removing Organic Micropollutants? *Water Science and Technology* 57, 251–255. doi:10.2166/wst.2008.825

Lerer, S.M., Arnbjerg-Nielsen, K. and Mikkelsen, P.S. 2015. A Mapping of Tools for Informing Water Sensitive Urban Design Planning Decisions – Questions, Aspects and Context Sensitivity. *Water* 7, 993–1012. doi:10.3390/w7030993

Makropoulos, C.K., Natsis, K., Liu, S., Mittas, K. and Butler, D. 2008. Decision Support for Sustainable Option Selection in Integrated Urban Water Management. *Environmental Modelling & Software* 23, 1448–1460. doi:10.1016/j.envsoft.2008.04.010

Marseille Statement. The UNESCO Symposium on Frontiers in Urban Water Management: Deadlock or Hope? 2001. *Urban Water* 3, 129–130. doi:10.1016/S1462–0758(01)00051–00056

McHarg, I.L. 1995. *Design with Nature*. John Wiley & Sons, New York.

Miller-Robbie, L., Ramaswami, A. and Amerasinghe, P. 2017. Wastewater Treatment and Reuse in Urban Agriculture: Exploring the Food, Energy, Water, and Health Nexus in Hyderabad, India. *Environmental Research Letters* 12, 75005. doi:10.1088/1748–9326/aa6bfe

Mitchell, V. 2006. Applying Integrated Urban Water Management Concepts: A Review of Australian Experience. *Environmental Management* 37, 589–605. doi:10.1007/s00267-004-0252-1

Mollinga, P.P., Dixit, A. and Athukorala, K. 2006. *Integrated Water Resources Management: Global Theory, Emerging Practice and Local Needs*. Sage, New Delhi, Thousand Oaks and London.

Novotny, V., Ahern, J. and Brown, P. 2010. *Water Centric Sustainable Communities*. John Wiley and Sons, Hoboken.

Sapkota, M., Arora, M., Malano, H., Moglia, M., Sharma, A., George, B. and Pamminger, F. 2014. An Overview of Hybrid Water Supply Systems in the Context of Urban Water Management: Challenges and Opportunities. *Water* 7, 153–174. doi:10.3390/w7010153

UN-HABITAT. 2003. *Water and Sanitation in the World's Cities.* Earthscan, London and Sterling, VA.

United Nations. 2015. *Sustainable Development Goals* [WWW Document]. Sustainable Development Knowledge Platform, Department of Economic and Social Affairs. https://sustainabledevelopment.un.org/?menu=1300.

United Nations Economic and Social Council. 1997. *Comprehensive Assessment of the Freshwater Resources of the World* (Report to the Secretary General No. E/CN.17/1997/9).

United Nations General Assembly. 2000. *United Nations Millennium Declaration* (Resolution adopted by the General Assembly No. A/RES/55/2).

United Nations General Assembly. 2010. *The Human Right to Water and Sanitation* (Resolution adopted by the General Assembly No. A/RES/64/292). United Nations, New York.

United Nations Sustainable Development. 1992. *Agenda 21*, in: United Nations Conference on Environment and Development. Presented at the United Nations Conference on Sustainable Development, Rio De Janeiro, Brazil.

United Nations World Water Assessment Programme. 2015. *The United Nations World Water Development Report 2015: Water for a Sustainable World.* UNSECO, Paris.

WCED. 1987. Towards Sustainable Development, in: World Commission on Environment and Development (Ed.), *Our Common Future.* Oxford University Press, Oxford and New York, pp. 43–65.

Wong, T. and Brown, R. 2009. The Water Sensitive City: Principles for Practice. *Water Science and Technology* 60(3), 673–682.

Zeitoun, M., Lankford, B., Krueger, T., Forsyth, T., Carter, R., Hoekstra, A.Y., Taylor, R., Varis, O., Cleaver, F., Boelens, R., Swatuk, L., Tickner, D., Scott, C.A., Mirumachi, N. and Matthews, N. 2016. Reductionist and Integrative Research Approaches to Complex Water Security Policy Challenges. *Global Environmental Change* 39, 143–154. doi:10.1016/j.gloenvcha.2016.04.010

3   Constructing infrastructure

Introduction

Infrastructures are large-scale urban technical systems. They comprise the basic technological fabric of the city, delivering essential urban functions. Infrastructures are at the same time intensely social and political and inseparable from the economy and everyday life of the city. The technologies of infrastructure exist within a broader social and political context, which in turn depend upon infrastructure to function, and so human interests and technical objects become entwined. Infrastructure systems provide a meeting point for the physical laws of science and engineering and humanly defined laws of politics, ethics, economics and society. How we understand the relationship between technology, values, politics and society has important implications for design, policy and operation of infrastructure.

Conventional engineering theories of infrastructure largely conform to an instrumentalist approach to technology, which understands technology as humanly controlled and value neutral (Feenberg, 1993). Under this view, infrastructure is an outcome of human planning and design on the basis of objective, technical rationality, and it provides a neutral background upon which culture, society, economy and politics develop. By contrast, critical accounts of infrastructure highlight the mutual influence of political and social context, and of infrastructure planning and engineering design (Feenberg, 1993). Philosopher Andrew Feenberg's (1993, p. 81) critical theory of technology draws attention to the political and social values that are intrinsic to modern technologies, industrialisation and infrastructure:

> Social purposes are 'embodied' in the technology and are not therefore mere extrinsic ends to which a neutral tool might be put. The embodiment of specific purposes is achieved through the 'fit' of the technology and its social environment.

Infrastructure systems embody social and political values and shape urban possibilities. Critical theories of technology and infrastructure draw empirical and methodological support from social constructivist accounts of science and technology, which show the influence of social processes on the production of scientific knowledge and technical artefacts (see for example Bijker et al., 1989; Bijker, 1997; Latour and Woolgar, 1979; Wajcman, 1991).

Infrastructures for energy, communications, transport, water and waste underpin the basic functioning of modern cities. They mediate relationships between cities and the natural environment, providing a buffer against environmental change and continuous access to basic material resources. Where they function effectively, infrastructures enable people to live in cities free from the cycles and extremes of the natural environment. Energy systems provide heating, cooling and lighting that support relatively stable and comfortable conditions for living and working, whatever the season or time of day. Drainage systems protect streets and homes from inundation during most rainstorms, and water and sanitation infrastructure operate consistently and continuously in all but the most extreme droughts. Infrastructures provide an essential coherence to modern life and stabilise relationships between people and nature (Edwards, 2003). In providing water, energy, mobility and waste removal wherever and whenever it is needed in cities, infrastructures facilitate modern lifestyles that are so characteristically urban. This has had tremendous benefits for human development, but as populations and consumption continue to grow, impacts on the natural environment through resource depletion and pollution have become more apparent.

This chapter provides a brief introduction to different trends and perspectives on infrastructure to emphasise the complexity of interactions between social, political, economic, environmental and technical elements in these important urban systems. It begins by pointing out the invisibility of most effective infrastructures, and their role as background to everyday life in cities. Trends away from public investment and the increasing role of the private sector are reviewed, and the technical and social movements towards decentralisation are explored. Gender is recognised as a key factor in integrated water management and sustainable development, and the chapter ends with some critical feminist perspectives on infrastructure and cities. This is not intended as a comprehensive analysis of the complex relationship between infrastructure, cities and nature, but to illustrate the need for technical, social, ecological and political issues to be considered together in deciding upon pathways to sustainability.

Urban background

Infrastructural technologies are largely invisible in cities in the developed world. Infrastructure and its services and resources form part of the background of everyday life, only entering the users' consciousness when something breaks down or when resources are scarce or absent altogether (Edwards, 2003). It is the absence of infrastructure in cities of the Global South that is noteworthy, not usually their role in shaping rates of resource consumption and lifestyles in cities where they function effectively.

Infrastructure systems are central in structuring modern patterns of consumption of natural resources. In contrast to other forms of consumption, consumption of resources through infrastructure services is inconspicuous, largely unnoticed (Shove, 2004). For example, the flushing toilet, the ubiquitous device at the centre of Western sanitation infrastructure, is only possible because of a continuous supply of water and continuous disposal of waste through sewers. The toilet is a node in vast water and wastewater infrastructure networks. Collectively, toilets across a city have significant impacts on water resources and the environment, and yet users barely register the freshwater that is polluted and the energy that is consumed in pumping and treatment associated with every flush. Water is visible in the flushing toilet, yet water is not seen as a resource or an element of nature. Consumption of a 600-millilitre bottle of water may be a conspicuous expression of health consciousness associated with a particular lifestyle and subject to environmental debate, but consumption of 6 litres of water in flushing the toilet is mostly unconscious and uncontested.

Even in circumstances where resource consumption is of concern, such as during droughts, infrastructure itself is usually spared scrutiny as resources and consumer behaviour, the two ends of the pipe, are subject to increased control and public debate (Bell, 2009). Drought discourse focusses on the need for consumers to change their behaviour, and for water resource managers to develop new resources, but rarely addresses the relationship between the two. Consumer behaviour has evolved within an infrastructural model of endless supply and continual expansion. Drought provides a temporary disruption to this assumption, and consumers have shown strong capacity to adapt behaviour accordingly. Drought has also prompted investment in desalination and dams to secure resources, to avoid future vulnerabilities (Kaika, 2003; Turner et al., 2016). However, in order to achieve longer-term changes in patterns of consumption and cultural expectations it is likely that the model of infrastructure that links consumer behaviour to water resources will also need to be reconsidered.

Infrastructure embodies values, shaping access and inclusion to urban life and citizenship (Winner, 1988). Access to infrastructure services of water, sanitation, energy, transport and data is widely understood to be a basic need and right of modern citizens. Water, energy and communications infrastructure also impact on the ability of different groups to participate in public and social life. Access to infrastructure services can improve health and wellbeing, and reduce daily work to find alternatives, enabling more time and attention to be devoted to work, education and social life. The difficulties faced by people with disabilities, the elderly, and parents and carers in accessing public transport, and the changes to provision in recent decades, most clearly demonstrate the impact of access to infrastructure on participation in society. In the Global South people living in informal settlements and rural areas are particularly likely to be cut off from formalised infrastructure services, and delivery of those services is a key objective for pro-poor and human rights–based development programmes. In the Global North crises such as drinking water contamination in the US city of Flint and the 'eat or heat' decisions of those living in fuel poverty demonstrate that access to basic infrastructure services for all people in society remains challenging. Affordable and equitable access to infrastructure is well established as a basic need for modern citizens and in the case of water and sanitation is now recognised as a human right, but meeting these obligations remains challenging in many parts of the world.

Ownership

For most of the twentieth century, provision of universal access to basic infrastructure was a goal of municipal and national governments (Marvin and Graham, 2001). Access to water and sewerage was deemed essential to good public health and a basic expectation of modern citizenship. As such, the state had a major role to play in the provision and management of infrastructure, which was seen to be a public good. Public investment in infrastructure was thought to underpin economic and social development. Infrastructure networks spread across cities and countries, linking citizens to the benefits of modern technology and resources. Such investment depended on governments having ready access to finance to fund the construction and ongoing revenue to fund operation and maintenance.

Since the 1980s changes in public policy associated with neoliberal market liberalisation and privatisation have had profound implications for infrastructure provision around the world. Private sector involvement in infrastructure provision expanded considerably in the 30 years prior to the financial crisis of the early twenty-first century. Infrastructure investment came to be seen as a key economic stimulus measure, but often in partnership with private investors and providers rather than through purely public ownership and operation. The move away from state ownership and the goal of universal provision towards a growing role for the private sector in infrastructure systems at the end of the twentieth century is documented in Marvin and Graham's (2001) *Splintering Urbanism*. Their work highlights the social, environmental and economic inequalities that can become entrenched when access to infrastructure services is based on profitability rather than universality. In many cities it is now possible to observe well-connected, middle- and upper-class suburbs and central business districts with world-class access to water and other infrastructure services, side by side with slum settlements that have no formal provision. The challenge of providing basic services to the urban poor is made more complex by the often unplanned, illegal status of slum settlements and the difficulty of obtaining finance to build or extend infrastructure systems to low-income, less profitable areas. As a result, the poorest city residents may be forced to pay the highest prices for water, resorting to private or illegal vendors who charge higher rates than the centralised utility (Allen et al., 2006; Swyngedouw, 2004).

In the years following the 2008 financial crisis and global recession, politicians in the Global North have shown renewed interest in infrastructure as an economic stimulus measure, creating jobs and improving local economic conditions. Lack of investment in infrastructure in the late twentieth century contributed to the decline in conditions and function, leaving some infrastructure unsafe, inefficient and unable to adequately serve growing populations and expectations in the twenty-first century. Engineering institutions such as the American Society of Civil Engineers and the UK-based Institution of Civil Engineers have produced 'report card'–style analyses of national infrastructure (ASCE, 2017; ICE, 2014). Governments in both countries have received notably low grades from their engineers, as the means of highlighting the 'crisis' of infrastructure and the need for urgent investment. With poor-quality infrastructure, economic growth and public safety are threatened. Renewed investment in infrastructure is a source of both economic activity and growth in itself, and provides the basis for stable long-term investment and economic development. Governments stimulate investment in infrastructure through direct public borrowing and spending on specific projects, through government guarantees of private sector investment, and through the creation of a range of policies, subsidies and financial mechanisms to facilitate private investment.

Analysis of the 'financialisation' of infrastructure, referring to the increasing interest of the finance sector in complex, profitable arrangements for funding infrastructure investment, shows the influence of banks, pension funds and other private investors in shaping the feasibility of different infrastructure options. For instance, Loftus and March (2016) contend that financial interests have been significant in decisions to build desalination plants in Spain and England. Large capital projects tend to be more attractive to financial institutions than decentralised or demand-side solutions to and resource scarcity, hence they are more prevalent in policy and investment decisions.

Decentralisation

Engineering and policy discourse about the relationship between infrastructure and the economy tends to focus on large infrastructure projects and networks. Attention to renewal and expansion of existing systems implies that models of infrastructure provision that were established in the nineteenth and twentieth centuries, based on centralised provision and large, interconnected systems, are essentially sound. The technical form of infrastructure is rarely questioned in policy and engineering discourse, beyond opportunities for integrating digital technologies and smart metering and control. Decentralised systems such as local renewable energy schemes and household- or building-scale water recycling tend not to feature prominently in engineering and government analyses of the infrastructure crises and strategies for infrastructure-led economic stimulus and urban regeneration, yet are often favoured by environmentalists.

The failure of top-down models of development, based on direct transfer of industrial and technological systems from the Global North to the Global South, to deliver improvements in the lives of the poorest people provided impetus for the emergence of the 'appropriate technology' movement from the 1960s until the 1980s (Smith et al., 2016). Partly inspired by ideas presented in E.F. Schumacher's (1973) classic *Small is Beautiful*, calls for appropriate technology challenged the transfer of large-scale industrial development to the developing world and called for development that was rooted in local environments and cultures and operated on a 'human scale'. Appropriate technologies meet local needs, are able to be fabricated and maintained using local resources and skills, and have net positive impacts on the local environment. The principles of appropriate technology, also known as 'alternative technology' or 'intermediate technology', also found relevance in the Global North, particularly amongst environmentalists. Decentralisation of infrastructure, and associated political and social institutions, has become a common theme in much environmental literature, activism and design. The alternative technology movement in industrialised economies promoted energy efficiency, renewable energy, composting, rainwater harvesting, greywater reuse and other alternatives to centralised infrastructural systems.

Recently, the aims of the appropriate technology movement have been reformulated and adapted to become known as 'grassroots innovation' (Smith et al., 2016). Grassroots innovation refers to a range of initiatives in different countries, including the Honey Bee Networks in India, technologies for social inclusion in Latin America and, to some extent, hacker and maker spaces in the Global North. In contrast to the appropriate technology movement, grassroots innovation places a stronger emphasis on local participation in the processes of innovation and design, rather than relying on experts or engineers to develop technologies that are suited to local needs. Social innovation is therefore as significant as technical innovation, emphasising inclusion in developing and delivering technologies from the bottom up.

Grassroots innovation and appropriate technology have delivered decentralised infrastructure services such as renewable energy, waste management and recycling, and water and wastewater services. They can also highlight the failures of top-down development processes to deliver basic infrastructure services to local people, particularly poor and marginalised communities. They also represent an alternative pathway to sustainable development, highlighting the need for social as well as technological change.

Decentralisation of technology and infrastructure is a common theme in ecological communities and environmental activism. However, decentralised technology is not necessarily more ecologically sound or democratic.

Centralised systems lose resources in distribution, but may deliver economies of scale that mean they are more resource and economically efficient than decentralised systems. Decentralised technology is popular with 'off-gridders' of all political persuasions, including the wealthy who do not want to rely on shared public services, libertarians who are suspicious of centralised authority, and environmentalists who want to live within local resource limits. Philosopher Langdon Winner has pointed out that decentralised technology may also imply decentralisation of social, economic and political organisation, though this could be to capitalist, socialist or anarchist ends (Winner, 1988).

The scale of technology need not be the same as the scale of ownership, institutional regulation and management. Van Vliet et al. (2005) analyse the role of infrastructure in shaping patterns of consumption, and the potential for alternative configurations of both technologies and institutions, including various arrangements which cut through the conventional dichotomy between centralised and decentralised systems. They reveal a move from centralised infrastructure provision and explore the potential for differentiation in infrastructure provision to lead to more sustainable modes of provision. They demonstrate that the scale of management need not match the scale of technologies and service networks. It is possible for small-scale technologies to be managed by large-scale institutions, and small-scale institutions can act as providers and brokers for larger networks of provision.

Infrastructure features prominently in future cities discourse as an opportunity for social and technical innovation (Zanella et al., 2014). Digital infrastructure and its integration with existing infrastructure are presented as 'connecting' society, government and the economy. Sensor networks provide data about the operation of infrastructure systems, including demand, and developments in control systems allow for improved system management and operation. Smart grids for electricity allow integration of renewable energy and sources, controlling electricity production to meet demand and allowing for responsive, variable tariffs and wholesale electricity prices to even out demand and supply. Smart traffic management systems control signals and speed limits and open and close lanes to manage road traffic flows. Partial or full automation of rail has improved efficiency and safety of services. Smart meters and home systems provide consumers with more information about their resource use and enable remote control of household devices and systems (Boyle et al., 2013; Depuru et al., 2011). Data about consumer use of infrastructure services provides new opportunities to analyse demand to identify opportunities to reduce it and to plan future infrastructure services to meet demand more efficiently and to improve services standards. ICT and smart city technologies also provide opportunities for centralised control of decentralised systems, allowing operational efficiency and reliability without the distribution inefficiencies associated with centralised infrastructure networks.

Digital technologies and innovative manufacturing present opportunities for more decentralised infrastructure systems, matching supply and demand instantaneously, and remotely controlling household technologies and infrastructure systems. Technical innovation and diffusion are driven or constrained by social and institutional change, proposing wider-ranging transformation than conventional 'infrastructure renewal' schemes. Changing values and social structures drive different forms of infrastructure, and new technologies enable different ways of living, working, socialising and building communities. Local economic activity, new business models facilitated by digital technologies, social enterprises and knowledge-based industries provide a different account of 'the economy' and its role in supporting sustainable communities and prosperity.

Gender

Gender is a significant factor in shaping infrastructural relationships. It has been recognised in IWRM and development processes as an important element in infrastructure sustainability. Practically, this means ensuring women are represented in participation processes and women's needs are accounted for in design and decision-making. Beyond the obvious need to include women in development processes, gender and infrastructure have deeper relationships that reflect wider cultural assumptions and associations.

Infrastructure is typically associated with masculine constructions of technology and engineering (Wajcman, 1991). Masculine association with infrastructure as technology include an obsession with hierarchical control and the pre-eminence of 'hard' technical solutions over 'soft' social processes. Whilst engineering work, like gender, rarely conforms to such stereotypical binaries, the social construction of gender, engineering and infrastructure is shaped by persistent cultural associations between women and nature, and men and technology.

Even in the early decades of the twenty-first century the engineering profession remains dominated by men, as does the management of water utilities. This is despite higher than average participation by women in engineering disciplines associated with water infrastructure, such as chemical, civil and environmental engineering. Notwithstanding genderless needs for bathing and toilet flushing, the private use of water remains largely the domain of women. This is consistent with nineteenth-century divisions of labour, urban space and governance. Whilst women have made considerable inroads into other areas of public life, and gender relationships have undergone significant transformations, water infrastructure remains a largely masculine domain. The consumption of water is associated with private, feminine activities and spaces, and the production of water remains a highly technical, masculine field of action.

Feminist urban scholarship uncovers the complexity of gender relationships and power in cities, including constructions of public and private space (Watson, 2002). Feminists have shown the problematic constructions of gender within the modern city, particularly in relation to women's capacity to engage in urban public life. Architecture and urban planning have orchestrated the separation between women and men, private and public, home and paid employment, consumption and production, suburb and city. While people do not actually live according to these dichotomies, these deeply rooted cultural associations continue to influence decisions and have an impact on women's lives (Wajcman, 1991).

The provision of clean water and sewerage services to homes via urban infrastructure systems has very important benefits for women. Piped water reduces the burden of sourcing and carting water, hot water systems reduce effort spent heating water and carting it within the home, and improved standards of cleanliness and hygiene reduce illness and the work of caring for the sick. However, water systems have also created new work in the home associated with ever-higher cultural standards of cleanliness, which has largely been taken up by women (Cowan, 1983).

The modern city itself provides opportunities and freedoms for women. Elizabeth Wilson's (1992) *The Sphinx in the City* provides an account of women's freedom and discipline in cities including London, Paris, New York, Chicago, Sao Paolo and Accra. She shows how changes in urban governance, culture and form impact on women's particular experience of the city. For instance, the sanitary movement in nineteenth-century London led to the provision of water and sewerage infrastructure, which undoubtedly improved women's lives but was also associated with a culture of urban control and discipline which limited women's ability to engage in public life and paid work.

Pointing out an association between masculine domination of water in public, engineering-led utilities and the feminine, private experience of everyday water is not to blame men for water shortages or to suggest that water consumption would be any different if they did more housework. Nor is it meant to imply that women engineers would design or manage infrastructure systems more sustainably. It simply demonstrates the patterns of domination and control, of both water and women, reflected in the systems and spaces of the modern city.

Infrastructure and sustainable cities

Infrastructures are large technical systems that shape citizenship, participation in society, and the everyday activities of modern lifestyles. They both enable and constrain particular behaviours and cultures, stabilising social and political relationships and mediating relationships between people and the environment. Infrastructures and their role in society are largely invisible and taken for granted, and they are only subject to debate and controversy when they fail or are absent.

Nature, gender, technology, culture, politics and society all intertwine in cities and are subject to constant change. The sustainability of urban systems is an outcome of these complex networks of relationships. Patterns of resource use are shaped by patterns of social relationships, infrastructure and technology. Moving towards more sustainable water use and drainage in cities requires knowledge of how these different factors and trends influence current water systems and proposals for new technologies and infrastructures.

References

Allen, A., Davila, J.D. and Hofmann, P. 2006. The Peri-Urban Water Poor: Citizens or Consumers? *Environment and Urbanization* 18, 333–351. doi:10.1177/0956247806069608

ASCE. 2017. *ASCE's 2017 Infrastructure Report Card | GPA: D+* [WWW Document]. ASCE's 2017 Infrastructure Report Card. www.infrastructurereportcard.org.

Bell, S. 2009. The Driest Continent and the Greediest Water Company: Newspaper Reporting of Drought in Sydney and London. *International Journal of Environmental Studies* 66, 581–589. doi:10.1080/00207230903239220

Bijker, W.E. 1997. *Of Bicycles, Bakelites, and Bulbs: Toward a Theory of Sociotechnical Change*. MIT Press, Cambridge, MA.

Bijker, W.E., Hughes, T.P. and Pinch, T. 1989. *The Social Construction of Technological Systems: New Directions in the Sociology and History of Technology*. MIT Press, Cambridge, MA.

Boyle, T., Giurco, D., Mukheibir, P., Liu, A., Moy, C., White, S. and Stewart, R. 2013. Intelligent Metering for Urban Water: A Review. *Water* 5, 1052–1081. doi:10.3390/w5031052

Cowan, R.S. 1983. *More Work for Mother*. Basic Books, New York.

Depuru, S.S.S.R., Wang, L. and Devabhaktuni, V. 2011. Smart Meters for Power Grid: Challenges, Issues, Advantages and Status. *Renewable and Sustainable Energy Reviews* 15, 2736–2742. doi:10.1016/j.rser.2011.02.039

Edwards, P. 2003. Infrastructure and Modernity: Force, Time and Social Organization in the History of Sociotechnical Systems, in: Misa, T., Brey, P. and Feenberg, A. (Eds.), *Modernity and Technology*. MIT Press, Cambridge, MA.

Feenberg, A. 1993. *Critical Theory of Technology*. Oxford University Press, Oxford.

ICE. 2014. *State of the Nation: Infrastructure 2014*. Institution of Civil Engineers, London.

Kaika, M. 2003. Constructing Scarcity and Sensationalising Water Politics: 170 Days That Shook Athens. *Antipode* 35, 919–954. doi:10.1111/j.1467–8330.2003.00365.x

Latour, B. and Woolgar, S. 1979. *Laboratory Life*. Sage, London.

Loftus, A. and March, H. 2016. Financializing Desalination: Rethinking the Returns of Big Infrastructure. *International Journal of Urban and Regional Research* 40, 46–61. doi:10.1111/1468–2427.12342

Marvin, S. and Graham, S. 2001. *Splintering Urbanism*. Routledge, London.

Schumacher, E.F. 1973. *Small Is Beautiful*. ABACUS, London.

Shove, E. 2004. *Comfort, Cleanliness and Convenience: The Social Organization of Normality*. Berg Publishers, Oxford.

Smith, A., Fressoli, M., Abrol, D., Arond, E. and Ely, A. 2016. *Grassroots Innovation Movements*. Routledge, Abingdon, Oxon and New York.

Swyngedouw, E. 2004. *Social Power and the Urbanization of Water: Flows of Power*. Oxford University Press.

Turner, A., White, S., Chong, J., Dickinson, M., Cooley, H. and Donnelley, K. 2016. *Managing Drought: Learning from Australia*. Pacific Institute, Oakland.

Vliet, B.V., Shove, E. and Chappells, H. 2005. *Infrastructures of Consumption: Environmental Innovation in the Utility Industries*. Earthscan, London and Sterling, VA.

Wajcman, J. 1991. *Feminism Confronts Technology*. Penn State Press, University Park, PA.

Watson, S. 2002. City A/genders, in: Bridge, G. and Watson, S. (Eds.), Wiley-Blackwell, Malden, pp. 290–296.

Wilson, E. 1992. *The Sphinx in the City: Urban Life, the Control of Disorder, and Women*. University of California Press, Berkeley.

Winner, L. 1988. *The Whale and the Reactor: A Search for Limits in an Age of High Technology*. University of Chicago Press, Chicago.

Zanella, A., Bui, N., Castellani, A., Vangelista, L. and Zorzi, M. 2014. Internet of Things for Smart Cities. *IEEE Internet of Things Journal* 1, 22–32. doi:10.1109/JIOT.2014.2306328

4 Framing cities and nature

Introduction

Water in cities defies categorisation into purely technical, natural, social or political components. Water infrastructure is the subject of competing meanings and narratives, which frame relationships between cities and nature in different terms. Understanding the underlying assumptions in how different versions of sustainability are constructed helps to identify political, theoretical and philosophical positions represented within debates about water and technologies in cities. Technology is not merely a neutral tool at the service of different political narratives, but plays an active role in how relationships between cities and nature are constructed. If we are to understand how technology can contribute to or undermine urban sustainability, it is useful to understand how it is discussed within alternative conceptual and discursive frameworks.

A framework provides a structure on which to build an understanding of water in cities, in order to understand the role of technology. It includes systems of knowledge and discourse. Conceptual frameworks provide a mental structure from which to analyse our experience of the world. According to philosopher Karen Warren a conceptual framework is

> a set of *basic* beliefs, values, attitudes, and assumptions which shape and reflect how one views oneself and one's world. It is a socially constructed lens through which we perceive ourselves and others. It is affected by such factors as gender, race, class, age, affectional orientation, nationality and religious background.
>
> (Warren, 1990, p. 127)

Discourse analysis has been used to analyse the terms on which the environment is discussed in policy, the media and public life (Darier, 1999; Hajer, 1995; Myerson and Rydin, 1996). According to John Dryzek, a discourse is

> a shared way of apprehending the world. . . . Discourses construct meanings and relationships, helping define common sense and legitimate

knowledge. Each discourse rests on assumptions, judgements, and contentions that provide the basic terms for analysis, debates, agreements and disagreements.

(Dryzek, 1997, p. 9)

Dryzek analysed environmental politics in the 1990s, identifying nine overlapping but distinct discourses – survivalism, prometheanism, administrative rationalism, democratic pragmatism, economic rationalism, sustainable development, ecological modernisation, green consciousness and green politics (Dryzek, 1997). Water and infrastructure may feature in each of the nine environmental discourses identified by Dryzek, but for the purpose of this book, most discussions of urban water sustainability fall within five conceptual frameworks – sustainable development, ecological modernisation, socio-technical systems, political ecology and radical ecology.

A framework is not merely an individual lens through which to view the world, but the foundation for shared narratives, meanings and knowledge. It shapes what is possible and how to achieve positive change in the world. This chapter outlines five frameworks for understanding urban water sustainability that are evident in research, practice, policy and activism. Sustainable development is the familiar framing of the need to deliver the benefits of development to the global population within ecological and resource limits. Ecological modernisation promises that environmental problems can be resolved by reforming the underlying institutions of modern society, with a central role for technological innovation. Socio-technical systems emphasise the co-evolution of technology and society in identifying opportunities to transition towards more sustainable infrastructure. Political ecology frames sustainability as a socio-environmental problem and highlights the connections between social inequality and ecological degradation. Radical ecology makes the case for fundamentally reconfiguring relationships between nature and society, emphasising the value of nature in its own right and as an active constituent of cities. The summary of the key features of each framework addresses historical development and core assumptions. The frameworks then form the basis for analysing the technological trends in urban water sustainability in Chapters 5–9.

Sustainable development

Sustainable development emerged in the 1980s as the global compromise between environmental concerns about population growth, resources and pollution, and the need for development to deliver the benefits of modern life to all people. Sustainable development aims to protect the environment but does not question underlying models of economic growth and industrialisation as the means to achieve benefits such as universal access to good health and education. As a global discourse it has always tried to balance the needs of countries at different stages of development, but recently sustainable development has become more prominent in relation to issues facing the Global South.

In the 1960s and '70s scientists such as Garrett Hardin and Paul Ehlich drew attention to the impacts of rapid growth of the human population and industrialisation, particularly in the Global South, and predicted catastrophic resource scarcity, pollution and environmental collapse if high growth rates were allowed to continue (Ehrlich, 1968; Hardin, 1968). During this period, the Limits to Growth Report, the concept of 'Spaceship Earth', the economics of Herman Daly and other contributions from scientists, activists and economists pointed out the fallacy of models of economics and development based on continuous growth in population and resource consumption on a finite planet (Daly, 1977; Meadows et al., 1972).

Patterns of development based on industrialisation not only depleted non-renewable resources at an unsustainable rate but also caused harmful environmental pollution. Rachel Carson's book *Silent Spring* drew attention to the environmental impacts of pesticides in the early 1960s (Carson, 1962). By the late 1970s environmental controversies in the US, such as the Three Mile Island nuclear accident and the Love Canal housing development built on a toxic waste dump, further demonstrated the risks to health and the environment of unconstrained industrial development. Optimism in the ability of science and technology to deliver unlimited benefits to humanity was widely questioned as the environmental and health consequences of industrialisation and economic growth were increasingly apparent.

The early environmental movement's emphasis on limiting growth and industrialisation was in conflict with strategies for economic development as the means for achieving universal access to basic health, education and a decent standard of living for the world's poorest people. Countries in the Global North had benefited from resource exploitation and pollution in earlier stages of their own development, and as colonial powers in the Global South. As the era of colonialism was ending, emerging nations and economies were unwilling to forgo their own opportunities for development to appease the concerns of northern environmental activists. Similarly, within developed countries, governments and industrialists were concerned that environmental protection should not require abandonment of the model of human progress based on continued economic growth. Economic growth and industrialisation were seen by some as the means to reducing population growth and protecting the environment, rather than simply the root cause of environmental harm and resource depletion.

Sustainable development promised to be the means by which to realise the benefits of development within environmental limits. The connection between development and environmental protection was recognised in the first UN Conference on the Human Environment in 1973 in Stockholm, and it was the focus of the World Commission on Environment and Development (WCED) chaired by Gro Harlem Brundtland. The 'Brundtland Report' of the WCED, *Our Common Future*, in 1988 made the famous definition of sustainable development as 'development that meets the needs of the present without compromising the ability of future generations to meet their own needs' (WCED, 1987). This definition confirmed the primacy of human development and poverty alleviation within sustainable development, and the importance of managing the environment for the long-term benefit of people, rather than for its own sake.

The 1992 'Earth Summit', the UN Conference on Environment and Development, in Rio de Janeiro affirmed international commitment to sustainable development and instigated various treaties on environmental protection, including the UN Framework Convention on Climate Change (UNFCCC). The final statement of the conference, 'Agenda 21', outlined the key issues for sustainable development and the means to address them at national and international levels (United Nations Sustainable Development, 1992). 'Local Agenda 21' came to characterise efforts by local governments to translate the principles of sustainable development to local planning and action, reflecting the 'think global, act local' slogan that was popularised at the time.

During the 1990s sustainable development was integrated into planning and policy by national and local governments around the world, and it has been characterised as the intersection between social, economic and environmental goals for development. Sustainable development has been variously represented as a Venn diagram, a three-legged stool, three pillars, and the triple bottom line of 'people, profit and planet' (Elkington, 1999). It represents efforts to integrate and reconcile equity, poverty alleviation and environmental protection with the mainstream goals of economic growth and development. While development and environmental protection were seen to be at odds in the early environmental movement, sustainable development proposed 'win-win-win' solutions, whereby development might deliver environmental and social improvements as well as economic growth.

In 2015 the 193 member countries of the General Assembly of the United Nations adopted the Sustainable Development Goals (SDGs), a set of targets to address poverty, economic growth, equality, environmental protection and climate change by 2030 (United Nations, 2015). The SDGs aim to be broader and more integrated than their predecessor the Millennium Development Goals (MDGs), which had been agreed in 2000 (United Nations General Assembly, 2000). The SDGs are universal and call for all countries to work to alleviate poverty and improve equality while reducing climate change and environmental impacts. Economic growth is recognised as underpinning poverty alleviation, improved health, education, job opportunities and social outcomes, and enabling environmental protection and climate change mitigation and adaptation.

Sustainable development has been influential in planning and policy in the Global North, but the Sustainable Development Goals of 2015 (SDGs) affirm a focus on poverty alleviation and delivering basic human services, which are of greatest concern to countries of the Global South.

The framework of sustainable development focusses on improving health, education, employment and human rights whilst at the same time protecting the environment and tackling climate change. Sustainable development emphasises the importance of gender equality and human rights, including for vulnerable populations such as children, the elderly and people living with disabilities. Economic growth is promoted as the means for achieving development and enabling protection of the environment from unsustainable exploitation and pollution. This framework underpins the concept of integrated water resources management, which emerged in global discussions through the Agenda 21 process. It also drives efforts to provide safe drinking water and sanitation, through the MDGs and SDGs.

Ecological modernisation

Ecological modernisation aims to move 'beyond apocalyptic orientations to see environmental problems as challenges for social, technical and economic reform, rather than as immutable consequences of industrialisation' (Mol and Sonnenfeld, 2000, p. 5). As a theory in environmental sociology and policy, ecological modernisation first emerged in the 1980s in Germany, the Netherlands and the UK. As with sustainable development, it developed in contrast to environmental theories of the 1970s which proposed that modernisation was the root cause of environmental harm, and that deindustrialisation and radical social change were the only solution for the ecological crisis.

Ecological modernisation theory and policy proposes more, not less, modern development as the solution to environmental problems. It holds that environmental damage and resource constraints can be solved by reforming modern institutions to incorporate ecological protection, with an important role for technological innovation, evidence-based policy and the market. By the turn of the twenty-first century, ecological modernisation had matured as a theory and had come to characterise dominant policy approaches to the environment in northern Europe and elsewhere. In 2000, Mol and Sonnenfeld outlined the core themes of the theory:

- Changing role of science and technology
- Increasing importance of market dynamics and economic agents
- Transformations in the role of the nation state
- Modifications in the position, role and ideology of social movements
- Changing discursive practices and emerging new ideologies

Science and technology play a central role in ecological modernisation, as the means to understand environmental harm and function and in driving technical innovation. New technologies have a central role in achieving sustainability through improved resource efficiency and reduced pollution. This includes improving the performance of existing technologies and transforming technical systems and infrastructures through constant innovation, driven by policy and markets.

Ecological modernisation has been aligned with neoliberal policies that emphasise the role of markets and individual consumer choice as the best way to achieve development and growth. However, it does not necessarily support neoliberal policy or capitalism as the only means to achieving sustainability. Unlike other environmental frameworks, ecological modernisation does not characterise capitalism and the market as an obstacle (Mol and Spaargaren, 2000). Capitalism can be reformed to account for the environment, and markets can be used to drive innovation and growth necessary to reduce pollution and improve resource efficiency. Market-based economics play a significant part in ecological modernisation policy: for example, the use of markets to allocate permits for water abstraction or pollutant discharge, and emphasis on the role of individual consumers in driving green production and consumption. Pricing is a core mechanism for allocation of scarce resources and a driver for improved efficiency.

The role of government shifts under ecological modernisation policy, with less emphasis on planning and regulation and more emphasis on partnership with the private sector to drive innovation and create the conditions for markets to deliver better environmental outcomes. Martin Hajer's (1995) study of air pollution policy in the UK and the Netherlands showed the influence of ecological modernisation, characterised as a shift towards enabling the private sector to innovate to reduce pollution rather than constraining development through regulation. Ecological modernisation also promotes more decentralised forms of government, within a broader conception of governance that recognises the role of the private sector and civil society alongside the state.

The environmental movement has had a contentious relationship with ecological modernisation theory and policy. Whilst some environmentalists maintain the need for more fundamental reorganisation of society and the economy, others have embraced ecological modernisation policies. Within ecological modernisation theory environmental and other social movements have a role to play as stakeholders and partners in devising policy to deliver environmental improvements and sustainable development, rather than an oppositional politics in which industrial development and environmentalism are at odds with one another.

Ecological modernisation sees the environment and sustainable development as core principles that transcend traditional ideological divisions. It represents the 'mainstreaming' of environmental issues into policy and politics, rather than environmentalism being a particular political movement or more aligned with radical politics. As modernisation and development are at the core of mainstream politics, ecological modernisation aims to bring the environment into mainstream political discourse, rather than it being a special interest.

Ecological modernisation policies are evident across the scale of water systems management, from market-based trading of water rights in catchments to household water metering. In order to make urban water systems sustainable in this framework, engineers need to develop technologies and systems that are economically and resource efficient. People in cities are conceived as individual customers of water and sanitation services. They respond as individuals to information about their consumption, particularly if that information is associated with a price signal.

While ecological modernisation theory and policy approaches developed in Europe, in the US similar ideas emerged around resource productivity, technological innovation and the reform of capitalism. The book *Natural Capitalism*, by Paul Hawken, Amory Lovins and Hunter Lovins, presented the ecological crisis as the foundation for the 'next industrial revolution' (Hawken et al., 2000). Natural capitalism rests on four core principles – radical resource productivity to dramatically improve the efficiency of technologies and manufacturing, biomimicry to design technologies inspired by natural systems, a service and flow economy decoupling economic growth from resource extraction, and investment in natural capital to restore ecosystems.

In 2015 the US-based Breakthrough Institute launched its 'Ecomodernist Manifesto', which promoted technological innovation as the key to resolving environmental problems (Asafu-Adjaye et al., 2015). Ecomodernism as described in the manifesto celebrates the achievements of modern development in alleviating poverty and proposes that further technical development, including urbanisation, agricultural intensification, desalination and nuclear energy, are the means for 'decoupling' human society from nature. Using technological systems to meet human needs reduces pressure on natural systems, creating space for nature to survive and be restored outside human settlement. The manifesto is specifically opposed to environmentalist calls for human society to live in harmony with nature, claiming that the only way for humans to develop and nature to survive is to decouple society from natural systems using technology.

The Ecomodernist Manifesto has some similarities to ecological modernisation theory, but its optimism regarding the role of technology in decoupling society and nature leaves it open to criticisms that have been raised since the 1980s. Ecological modernisation theory and policy has evolved since its early years to take better account of civil society and the role of culture in shaping everyday household practices that drive environmental resource consumption and impacts. Ecological modernisation theory also emphasises the need for reform of modern institutions such as the market, the state, civil society, science and technology to a greater extent than the ambitious and provocative manifesto.

Socio-technical systems

Infrastructure systems are both social and technical, as described in Chapter 3. Understanding water infrastructures as socio-technical systems that mediate relationships between people and their environment, enable and constrain particular forms of social life, and are governed by specific institutions and physical laws can help to reveal their role in stabilising or disrupting existing sustainable and unsustainable relationships between cities, people and nature. This presents a significant challenge to conventional disciplinary and theoretical perspectives under which social, technical and natural sciences have each evolved separately, with distinct knowledge systems, methods and languages. Understanding urban infrastructure and sustainability requires the ability to discuss social and technical issues at the same time, which is the aim of socio-technical systems frameworks of analysis. The specific analysis of 'socio-technical systems' as a framework for analysing urban infrastructure usually refers to theories and methods emerging from the sociology and anthropology of science and technology since the 1970s and '80s, some of which were discussed in Chapter 3. These include theories of large technical systems (LTS), the social construction of technology (SCOT), actor-network theory (ANT) and practice theories.

Large technical systems theories seek to understand the social forces that shape the formation and persistence of large technical systems such as electricity, communications and transport networks. In his landmark study of the history of electricity in the US, Germany and England, Thomas Hughes says that 'power systems are cultural artifacts' (Hughes, 1993, p. 2), and his study shows the political, economic and social factors that drove the development of local electricity supply systems from individual cities and isolated networks into interconnected regional grids.

Studies of the social construction of technology similarly aim to identify the social factors shaping technological artefacts and the 'sociotechnical ensemble' (Bijker, 1997). This theoretical tradition emerged partly as a critique of technological determinism, to demonstrate that technological development is the result of social and political negotiation and power rather than a neutral, technical rationality. SCOT analysis identifies the relevant social groups for any new technology and the different interpretations each group has of both the technology and the problem it poses to solve (Bijker et al., 1989; Bijker, 2009). As the technology develops, this 'interpretive flexibility' narrows, and the technology moves towards 'closure', as options for the form, function and meaning of the technology are reduced, and the final 'stable' artefact or system is confirmed. SCOT analysis highlights the social and political decisions and relationships that are embedded and taken for granted in these final, 'stabilised' technologies, despite their appearance as neutral, technical objects.

Actor-network theory emphasises the relationships between human and non-human actors that are built and stabilised in science, technology and society (Callon, 1986; Latour, 1993; Law, 1992). It emphasises the material or physical reality of these relations, and the agency of material artefacts as a balance to strongly 'social constructivist' perspectives. Rather than dividing reality into 'nature and culture' or 'science and society' as separate domains of knowledge and analysis, actor-network theory applies the same methods for understanding how networks of relationships between humans and non-humans are established and maintained. While social factors are important in shaping technologies, technological artefacts have their own physical reality, constraints and possibilities and are not mere 'social constructions'. Likewise, social life is shaped by, though not entirely determined by, scientific knowledge and technologies. This symmetry between human and non-human actors and attention to understanding how relationships between them are built or breakdown provides a framework for analysing socio-technical systems and opportunities for change.

In his book *We Have Never Been Modern* Latour shows that the modern idealised separation between nature and culture, science and society, and technology and values hides the work of 'hybrids' that mediate between these supposedly opposing domains of human thought and action (Latour, 1993). Infrastructure systems are classic 'hybrids' in Latour's terms – neither purely natural nor cultural, but a hybrid form that mediates between the two. Central to Latour's analysis is the idea that the modern world consists not of separate physical and social realms, but networks of relationships between human and non-human actors. Human values and social structures are not separate to the physical reality of nature and technology, but are built up through dynamic material relationships.

Susan Leigh Star (1999) defines infrastructure as a 'fundamentally relational concept, becoming real infrastructure only in relation to organized practice' (p. 380). Infrastructure exists only in the context of its use or as a problem to be solved. The water system consists of multiple meanings and uses, not simply pipes, pumps, valves, taps and treatment works. To the office worker beginning their day, water infrastructure is the system that is necessary to take a shower, to the plumber it is a system requiring repair, and to the city planner it is one variable in a complex urban planning process (Star, 1999). Infrastructure is all of these things – physical technologies and artefacts, daily habits and routines, expertise in design and maintenance, and strategic and political oversight in planning and management.

Urban and infrastructural change is constrained by existing networks and physical and institutional structures. Anique Hommels (2005) has written about the obduracy of urban socio-technical systems. Once cities are built, they are difficult to 'unbuild'. She outlines three conceptual models for understanding obduracy in cities – dominant frames that constrain professional and social groups' ability to see alternative possibilities; embeddedness of social and technical elements in close relationships; and persistent traditions that give rise to technical, institutional and professional path dependence. The obduracy of cities and infrastructures cannot be attributed to mere economics, politics or physical constraints, but arises from the complex interplay between different actors and factors.

Socio-technical theories used together with the sociological study of everyday life have shown the co-evolution of infrastructures, technologies and resource-using practices. Socio-technical studies of sustainable consumption emerged partly to fill a gap in ecological modernisation theory's focus on sustainable production, and to understand the complexity of relationships that shape consumption beyond a simplistic focus on technological efficiency and individual behaviour. Technology and behaviour shape each other in everyday life, creating habits and cultural expectations that drive resource consumption. Sociologist Elizabeth Shove has shown how technology, infrastructure and culture interact, leading to a 'ratcheting up' of consumption of energy and water (Shove, 2010, 2004). For instance, infrastructure systems provide a continuous supply of water and energy, enabling technologies such as automatic washing machines, which in turn drives personal and cultural expectations that clothes will be washed frequently, driving up demand for water and energy. While ecomodernist policy approaches focus on improving the efficiency of technologies, such as the washing machine, or changing behaviour, such as washing only a full load, a co-evolutionary approach points to wider drivers and constraints on resource-consuming practices.

Transitions theory has been an increasingly influential socio-technical approach in recent urban water sustainability research. Drawing on theories of complexity, innovation and socio-technical systems, transitions theory developed in the UK and Netherlands in the 2000s by researchers such as Rotmans, Grin and Gells (Geels, 2002; Geels and Schot, 2007; Smith et al., 2010). The 'multi-level perspective' addresses socio-technical systems change, and has been applied to energy systems, water and other infrastructures (Geels, 2011, 2010). This perspective promotes 'experimentation' as a way of initiating urban and socio-technical change within existing systems. Transitions theories have been used to frame urban water systems research, particularly in addressing the institutional barriers to innovation and implementation. Brown and others have shown the role of institutional experimentation as well as technical demonstration, and the importance of individual leadership and professional knowledge and expertise in urban water sustainability (Bos and Brown, 2012; Brown et al., 2013).

The socio-technical framing of urban water sustainability discusses water in terms of relationships between technology, culture, institutions, people and infrastructure. Water infrastructure, water use, water governance and the role of water in everyday life and urban culture interact and depend upon one another. Social and technical factors are analysed at the same time, addressing processes of innovation and cultural change rather than strict delineation between technological function and social life.

Political ecology

Urban political ecology frames sustainability as a fundamentally political question about how the costs and benefits of socio-environmental change are distributed (Loftus, 2009). Sustainability in this sense cannot be achieved without addressing underlying inequalities in society. It requires strong democratic engagement with socio-environmental problems and to construct solutions that enable more equitable and inclusive distribution of power and resources. This is distinct from conventional framings of sustainable development and ecological modernisation, which tend to depoliticise environmental problems. Political ecology questions the presumed social and political consensus of sustainability as the inevitable outcome of the reform of the existing political-economic system (Swyngedouw, 2009). It pays particular attention to social power relations, which determine who has access to environmental resources and who pays for the consequences of environmental degradation (Swyngedouw et al., 2002).

Political ecology begins with the recognition that social and ecological processes co-determine and co-constitute each other (Monstadt, 2009; Rodríguez-Labajos and Martínez-Alier, 2015). Cities are not 'unnatural', just as rural and 'wilderness' landscapes are not immune to social processes and cultural meanings. Social relations are shaped by ecological resources, the local environment and landscapes. Ecological processes are not just impacted by society, but landscapes and the environment itself are important in constructing social and cultural knowledge and meanings. Political ecology does not distinguish between social and cultural processes as distinct from biophysical nature, but refers to socio-ecological processes as environmental and social changes co-determine each other (Swyngedouw et al., 2002). Environmental change and conditions are the result of specific social, historical and political conditions and institutions. Social and political processes are likewise dependent upon the transformation of biophysical elements of the environment.

The co-evolution of social and environmental processes is captured in the political ecology concept of 'metabolism'. Drawing on Karl Marx's use of the word metabolism to describe the transformation of nature by labour, which was intensified by industrialisation and capitalism, in political ecology urban metabolism is used to analyse the circulation, transformation and exchange of resources that determines environmental and social conditions within and beyond the city. Urban metabolism in political ecology therefore draws attention to how the distribution and transformation of environmental resources and ecosystems is related to social, economic and political structures. Urban metabolism is also used in other sustainability analyses to focus on quantitative accounting and on modelling biophysical flows of materials in cities, without reference to society or politics. Urban political ecology draws attention to the uneven social and political processes and conditions that co-determine the flows and balances of materials and their ecological consequences, rather than conceiving of urban metabolism as a problem of resource efficiency and optimisation. For political ecology, urban metabolism has diverse, and sometime contradictory, social and political outcomes, with winners and losers in different places and social classes. For instance, improving living conditions based on higher water consumption within cities may depend upon declining environmental conditions due to water abstraction and pollution elsewhere (Swyngedouw et al., 2002).

Water infrastructure is a critical concern of urban political ecology, not only as a vital element of cities in itself, but also as an exemplar of the complex relationships and flows between nature, culture, technology, politics and bodies in cities. Political ecology frames urban water sustainability as a problem of uneven distribution of power and resources and the consequences of consumption and production in cities. Water and power are often found to flow in the same direction in cities. According to Erik Swyngedouw,

> Water (is) inherently political, and therefore contentious, and subject to all manner of tensions, conflicts and social struggles over its appropriation, transformation, and distribution, with socio-ecologically unevenly partitioned consequences.
>
> (Swyngedouw, 2015, p. 20)

Swyngedouw's (2004) account of the unequal access to water infrastructure in Guayaquil, Ecuador, describes the processes and structures of exclusion from basic public health infrastructure, including the problematic role of private capital and neoliberal policy agendas in water provision. Gandy's (2004) analysis of water infrastructure in Mumbai highlights similar factors at work in the continued failure of governments and private utility companies to deliver water to the urban poor, alongside the delivery of world-class services to wealthy districts in the city.

The changing role of the private sector in water infrastructure provision and the use of market-based policies and mechanisms to allocate water resources and manage water pollution are of particular concern to political ecology (Bakker, 2010, 2000). Water infrastructure in many cities in the UK, US and elsewhere was originally owned by private companies, and was transferred to municipal ownership in the late nineteenth and early twentieth centuries as water and sanitation provision became recognised as an important public health service requiring universal provision. Privatisation of the water sector has taken many forms since the 1970s, including increased use of private contractors to deliver water services, decentralised provision of water to buildings and residential developments outside the municipal network, and complete transfer of regional water infrastructure from public to private ownership.

Whether water infrastructure is under public or private ownership, since the 1970s water has become widely represented as a commodity, to be traded as an economic good, rather than as a public good or common resource (Loftus, 2009). The commodification of water has underpinned market-based trading of water rights and strategies for metering and water pricing as the means for managing water demand. It is also associated with the transformation of water utilities from public service institutions to more corporate, customer-focussed businesses.

The increasing role of the private sector at the end of the twentieth century corresponded to a general shift towards neoliberal politics, emphasising the market, competition, deregulation and individual choice as the means for improving efficiency, productivity and welfare. This was also associated with a shift away from the understanding role of government as the central actor in infrastructure planning, provision, investment and regulation, towards the concept of governance, which includes the private sector and civil society alongside government as agents responsible for delivering infrastructure or other services. In the UK and other countries, privatisation has required strong regulation of the private water sector to ensure affordability, maintain long-term investment and improve resilience and sustainability. In countries in the Global South with weaker governance arrangements and unstable governments, privatisation has failed to deliver promised improvements in access to the poor. When well regulated and scrutinised, the private sector can play an important role in delivering water and sanitation services, but privatisation itself is unlikely to lead to improved access, efficiency and sustainability (Bakker, 2010).

Urban political ecology analysis shows how these changes have increased the influence of private capital and the importance of profit making in delivering what was previously a public service, with potentially detrimental consequences for the environment, the urban poor, women and other marginalised social groups. The increased role of the private sector and importance of profitability, particularly in contexts where governments and civil society are underdeveloped, has reduced democratic accountability for infrastructure provision. Privatisation has also been associated with financialisation of infrastructure, which refers to the influence of financial investors seeking stable returns on capital investment as a driving force in infrastructure decisions (Loftus and March, 2016). Institutional investors in private water companies and other financial actors have an interest in promoting capital-intensive projects which create new capital assets that deliver a steady return, rather than wider consideration of options for improving sustainability.

Urban political ecologists have also demonstrated how the discourse of 'water crisis' in cities has been used as justification for privatisation, commodification and capital-intensive investments, without adequate analysis and deliberation about costs, benefits, environmental impacts and alternatives. Maria Kaika's (2006, 2003) analysis of water provision in Athens shows how the drought of 1989–1991 resulted in a decision to build a new dam and to corporatise the water utility, with particular benefits to private sector investors and engineering contractors, and without adequate consideration of the potential for demand management or other measures to improve overall sustainability. Karen Bakker's (2010) assessment of water privatisation around the world concludes that urban water crises are a result of poor management, rather than absolute water scarcity. Water scarcity is commonly represented as a resource crisis, either due to natural limits or climate change, yet water shortages and unequal access are the result of mismanagement of infrastructure and resources, highlighting the necessity to address water sustainability as a political as well as hydrological challenge.

Urban political ecology also provides analysis across different scales. Local processes of resource distribution and pollution connect to regional decisions about infrastructure and resources and are linked to national policies and international politics and investment. Urban water sustainability may not be a global crisis in terms of absolute water scarcity or environmental degradation, but global processes of deregulation, capital flows and trade have implications for socio-environmental and socio-hydrological conditions in cities, with uneven consequences for citizens.

Radical ecology

Emerging from the environmental movement in the 1970s and '80s, several distinct movements have developed within radical environmental philosophy and activism, each with a particular analysis and position on the underlying reasons for humanity's destruction of nature and propositions for alternative social and political structures. Deep ecology, social ecology and ecological feminism are three prominent schools of thought in environmental philosophy, politics and activism, each concerned with how to arrange societies in order to avoid human domination and destruction of nature.

Deep ecologists hold that the *anthropocentrism* (human-centred view) of dominant Western value systems is at the root of the ecological crisis. They call for an *ecocentric* system of values and society that places ecological concerns at the heart of all human culture and politics. Since human systems are part of ecological systems, our primary concern should be the preservation and maintenance of nature. Moreover, an ecocentric world view values nature *for its own sake*, in contrast to an anthropocentric view, which values nature only in terms of its use for humans. Non-human nature, particularly in wilderness areas, has a right to exist on its own terms, irrespective of potential economic or other value to humans. Deep ecology is closely associated with wilderness preservation, is generally anti-industrialist and supports strict control of human populations.

The term 'deep ecology' was first coined in 1972 in a paper by Norwegian philosopher Arne Naess (Naess, 1995). He contrasted 'deep ecology', based on a deep questioning of human relationships to nature, to 'shallow ecology', which characterises more conventional scientific and reformist approaches. Shallow ecology is anthropocentric, focusses primarily on pollution and resource depletion, and is ultimately concerned with the health and affluence of people in developed countries. Deep ecology involves a deep questioning of the goals and viability of industrial society, focusses on the interconnectedness of all life, and aims at restructuring society to achieve greater local autonomy and decentralisation.

Deep ecology has been criticised by others in the environmental movement for a lack of attention to the structure of society and its relationship to dominance of nature (Salleh, 1984). Its primary attention to the problems of anthropocentrism and its efforts to promote an ecocentric philosophy have been criticised for overlooking the relationships of domination and subjugation that occur within human society. Social ecologists and ecological feminists in particular have been critical of deep ecology, and their analyses and propositions for change draw on and extend existing critical frameworks to incorporate not only the oppression of nature but the oppression of the poor, women and others in society (Bookchin, 1984; Salleh, 1984).

The most prominent social ecologist of the twentieth century and clearest proponent of this branch of environmental philosophy was American Murray Bookchin. Bookchin's work, including his 1982 book *The Ecology of Freedom*, outlined key elements of social ecology, which is based on the proposition that the ecological crisis arises from deep-seated social problems due to the hierarchical structure of modern capitalist society (Bookchin, 2005). Human domination of nature is preceded by hierarchical relationships of domination and subjugation within industrial society. Social ecologists favour decentralised, local communities based on principles of self-organisation, which are in balance with local ecosystems as a means for reorganising society and resolving the ecological crisis. In contrast to deep ecology, social ecology places relationships of domination between humans at the centre of the ecological crisis. Drawing on long and varied traditions of anarchism, localism, decentralisation and the use of ecological examples and metaphors in designing human settlements and social structures, social ecology has had a significant influence on environmental politics and activism (Clark, 1998).

Ecological feminism (ecofeminism) identifies the cultural roots of the ecological crisis in the connection between the domination of women and the domination of nature (Merchant, 1989; Plumwood, 1993; Warren, 1990). *Critical ecofeminism* uses scholarly and political analysis to show that the domination of women and nature have the same historical and philosophical basis, and proposes that solutions to the ecological crisis must address relationships of domination within the Western cultural tradition, particularly gender relations (Plumwood, 1993; Warren, 1998). More controversially, *essentialist ecofeminism* draws direct parallels between women and nature, claiming that women are inherently closer to nature than men, and therefore suffer more directly when nature is dominated or destroyed (Mies and Shiva, 1993). Essentialist ecofeminism is rejected by many feminists for whom the direct association of women and nature re-enforces patriarchal systems of thought and stereotypes of women as more emotional and less rational than men (Hay, 2002).

Karen Warren (1990) outlined four minimal claims of ecofeminism:

- There is a connection between oppression of women and nature
- Linked oppression is sanctioned by a patriarchal framework

- Critique of patriarchy is grounded in ecological principles
- Ecological politics must be feminist

These demonstrate the need for ecological and feminist politics and action to be linked, as in fact they are both working to overcome the same structures of oppression. Although specific ecological and feminist issues may be distinct, they are linked through the same patterns of domination. Ecofeminist activism addresses a range of environmental issues, including urban, environmental justice and livelihood issues (Cuomo, 1998).

Framing urban water sustainability

The frameworks of urban water sustainability are distinct but overlapping, each providing a different emphasis for analysis and priorities for change. Table 4.1 summarises the key assumptions within each framework.

These frameworks are not paradigms that follow one after the other in a linear pathway. Nor are they fixed. They are dynamic social and political framings of the meaning of sustainability and how to achieve it in policy and practice. This is important for engineers, scientists, planners and activists who want to implement change through technology and science. These frameworks may be used in three modes – explanatory, normative and pragmatic.

The explanatory mode seeks to understand why things are the way they are. Why are some technologies preferred over others? Why are some elements of urban water sustainability emphasised in policy and public discourse while others are ignored? Why are some changes easy to implement while others seem impossible? The normative mode seeks to define how things should be. What are the barriers to implementing a particular vision of sustainability? How can policy be influenced to implement specific change? What should be valued most highly in implementing sustainability?

As a normative standpoint, each framework presents a vision for sustainability and how it should be achieved. They share an underlying concern to ensure water of sufficient quality and quantity for people and the environment, but they have different models of how to achieve this and how to manage costs and benefits to different social and ecological groups.

The pragmatic mode seeks to influence change within the constraints of how things are. This approach is more incremental and reformist than a normative definition of how things should be, and it is more active than mere explanation. Pragmatism in urban water sustainability may use particular frameworks strategically to influence change, but risks reinforcing dominant frameworks which may prevent more fundamental change. Pragmatic deployment of different framings also needs to maintain flexibility and adaptability, as frameworks and their relative influence and effectiveness in

Table 4.1 Frameworks for urban water sustainability

	Water	Technology	Politics	Society	Economy	Ecology
Sustainable development	Basic need	Appropriate	Internationalist	Poverty alleviation	Growth	For current and future generations
Ecological modernisation	Natural resource	Efficient	Neoliberal	Individuals	Market innovation	Ecosystem service
Socio-technical systems	Material	Co-evolves with society	Deliberative	Networks	Flows of materials and value	Shapes society and technology
Political ecology	Socio-ecological metabolic agent	Embodies socio-ecological relationships	Leftist	Co-constructed with ecology	Critical of global capitalism	Co-constructed with society
Radical ecology	Element of nature	Exploits nature	Ecocentric	Place-based community	Constrained by ecology	Valued intrinsically

delivering sustainability are far from fixed and are subject to social and political change.

The frameworks themselves are also dynamic. As evident in sustainable development, the discourse and knowledge within a given framework change over time. New framings emerge, such as political ecology, while others may be fading, such as radical ecology. This chapter has provided an overview of the most prominent framings in research, policy and activism at this point in time, but they are not fixed and may be incomplete. The purpose has been to identify frameworks from which to analyse the role of different technologies within urban water sustainability in the chapters to come.

References

Asafu-Adjaye, J., Blomqvist, L., Brand, S., Brook, B., Defries, R., Ellis, E., Foreman, C., Keith, D., Lewis, M., Mark, L., Nordhaus, T., Pielke Jr, R., Pritzker, R., Roy, J., Sagoff, M., Shellenberger, M., Stone, R. and Teague, P. 2015. *An Ecomodernist Manifesto*. www.ecomodernism.org

Bakker, K. 2010. *Privatizing Water*. Cornell University Press, Ithaca and London.

Bakker, K.J. 2000. Privatizing Water, Producing Scarcity: The Yorkshire Drought of 1995. *Economic Geography* 76, 4–27. doi:10.2307/144538

Bijker, W.E. 1997. *Of Bicycles, Bakelites, and Bulbs: Toward a Theory of Sociotechnical Change*. MIT Press, Cambridge, MA.

Bijker, W.E. 2009. Social Construction of Technology, in: Olsen, J.K.B., Pedersen, S.A. and Hendricks, V.F. (Eds.), *A Companion to the Philosophy of Technology*. Wiley-Blackwell, Chichester, pp. 88–94.

Bijker, W., Hughes, T.P. and Pinch, T. 1989. *The Social Construction of Technological Systems: New Directions in the Sociology and History of Technology*. MIT Press, Cambridge, MA.

Bookchin, M. 1984. Toward a Philosophy of Nature – The Bases for an Ecological Ethics, in: Tobias, M. (Ed.), *Deep Ecology*. Avant Books, San Francisco, CA, pp. 213–239.

Bookchin, M. 2005. *The Ecology of Freedom: The Emergence and Dissolution of Hierarchy*. AK Press, Oakland, CA.

Bos, J.J. and Brown, R.R. 2012. Governance Experimentation and Factors of Success in Socio-Technical Transitions in the Urban Water Sector. *Technological Forecasting and Social Change* 79, 1340–1353. doi:10.1016/j.techfore.2012.04.006

Brown, R.R., Farrelly, M.A. and Loorbach, D.A. 2013. Actors Working the Institutions in Sustainability Transitions: The Case of Melbourne's Stormwater Management. *Global Environmental Change* 23, 701–718. doi:10.1016/j.gloenvcha.2013.02.013

Callon, M. 1986. The Sociology of an Actor-Network: The Case of the Electric Vehicle, in: Callon, M., Law, J. and Rip, A. (Eds.), *Mapping the Dynamics of Science and Technology: Sociology of Science in the Real World*. Palgrave Macmillan, Basingstoke and London, pp. 19–34.

Carson, R. 1962. *Silent Spring*. Penguin Books, Harmondsworth.

Clark, J. 1998. A Social Ecology, in: *Environmental Philosophy*. Prentice-Hall Inc, Upper Saddle River, pp. 416–440.

Cuomo, C.J. 1998. *Feminism and Ecological Communities: An Ethic of Flourishing.* Routledge, New York.

Daly, H. 1977. *Steady-State Economics: The Economics of Biophysical Equilibrium and Moral Growth.* W. H. Freeman and Co., San Francisco.

Darier, E. 1999. Foucault and the Environment: An Introduction, in: Darier, E. (Ed.), *Discourses of the Environment.* Wiley-Blackwell, Oxford, pp. 1–33.

Dryzek, J.S. 1997. *The Politics of the Earth: Environmental Discourses.* Oxford University Press, Oxford and New York.

Ehrlich, P. 1968. *The Population Bomb.* Ballantine, New York.

Elkington, J. 1999. *Cannibals with Forks: The Triple Bottom Line of 21st Century Business.* Capstone Publishing Limited, Oxford.

Gandy, M. 2004. Rethinking Urban Metabolism: Water, Space and the Modern City. *City* 8, 363–379. doi:10.1080/1360481042000313509

Geels, F.W. 2002. Technological Transitions as Evolutionary Reconfiguration Processes: A Multi-Level Perspective and a Case-Study. *Research Policy, NELSON + WINTER + 20* 31, 1257–1274. doi:10.1016/S0048–7333(02)00062–00068

Geels, F.W. 2010. Ontologies, Socio-Technical Transitions (to Sustainability), and the Multi-Level Perspective. *Research Policy, Special Section on Innovation and Sustainability Transitions* 39, 495–510. doi:10.1016/j.respol.2010.01.022

Geels, F.W. 2011. The Multi-Level Perspective on Sustainability Transitions: Responses to Seven Criticisms. *Environmental Innovation and Societal Transitions* 1, 24–40. doi:10.1016/j.eist.2011.02.002

Geels, F.W. and Schot, J. 2007. Typology of Sociotechnical Transition Pathways. *Research Policy* 36, 399–417. doi:10.1016/j.respol.2007.01.003

Hajer, M. 1995. *The Politics of Environmental Discourse.* Oxford University Press, Oxford.

Hardin, G. 1968. The Tragedy of the Commons. *Science* 162, 1243–1248.

Hawken, P., Lovins, A. and Lovins, L.H. 2000. *Natural Capitalism.* Back Bay Books, New York.

Hay, P. 2002. *A Companion to Environmental Thought.* Edinburgh University Press, Edinburgh.

Hughes, T.P. 1993. *Networks of Power: Electrification in Western Society, 1880–1930.* Johns Hopkins University Press, Baltimore, MD.

Kaika, M. 2003. Constructing Scarcity and Sensationalising Water Politics: 170 Days That Shook Athens. *Antipode* 35, 919–954. doi:10.1111/j.1467–8330.2003.00365.x

Kaika, M. 2006. Dams as Symbols of Modernization: The Urbanization of Nature Between Geographical Imagination and Materiality. *Annals of the Association of American Geographers* 96, 276–301. doi:10.1111/j.1467–8306.2006.00478.x

Latour, B. 1993. *We Have Never Been Modern.* Pearson Education Ltd, Harlow, Essex.

Law, J. 1992. Notes on the Theory of the Actor-Network: Ordering, Strategy, and Heterogeneity. *Systemic Practice and Action Research* 5, 379–393.

Loftus, A. 2009. Rethinking Political Ecologies of Water. *Third World Quarterly* 30, 953–968. doi:10.1080/01436590902959198

Loftus, A. and March, H. 2016. Financializing Desalination: Rethinking the Returns of Big Infrastructure. *International Journal of Urban and Regional Research* 40, 46–61. doi:10.1111/1468–2427.12342

Meadows, D.H., Meadows, D.L., Randers, J. and Behrens, W.W. 1972. *Limits to Growth*. Universe Books, New York.

Merchant, C. 1989. *Ecological Revolutions: Nature, Gender, and Science in New England*. University of North Carolina Press, Chapel Hill.

Mies, M. and Shiva, V. 1993. *Ecofeminism*. Spinifex, Melbourne.

Mol, A.P.J. and Sonnenfeld, D.A. 2000. Ecological Modernisation Around the World: An Introduction. *Environmental Politics* 9, 1–14. doi:10.1080/09644010008414510

Mol, A.P.J. and Spaargaren, G. 2000. Ecological Modernisation Theory in Debate: A Review. *Environmental Politics* 9, 17–49. doi:10.1080/09644010008414511

Monstadt, J. 2009. Conceptualizing the Political Ecology of Urban Infrastructures: Insights from Technology and Urban Studies. *Environ Plan A* 41, 1924–1942. doi:10.1068/a4145

Myerson, G. and Rydin, Y. 1996. *The Language of Environment*. UCL Press, London.

Naess, A. 1995. The Deep Ecological Movement: Some philosophical aspects, in: Sessions, G. (Ed.), *Deep Ecology for the Twenty-First Century*. Shambala, Boston, pp. 151–155.

Plumwood, V. 1993. *Feminism and the Mastery of Nature*. Routledge, London.

Rodríguez-Labajos, B. and Martínez-Alier, J. 2015. Political Ecology of Water Conflicts. *WIREs Water* 2, 537–558. doi:10.1002/wat2.1092

Salleh, A.K. 1984. Deeper than Deep Ecology. *Environmental Ethics* 6, 339–345.

Shove, E. 2004. *Comfort, Cleanliness and Convenience: The Social Organization of Normality*. Berg Publishers, Oxford.

Shove, E. 2010. Beyond the ABC: Climate Change Policy and Theories of Social Change. *Environment and Planning A* 42, 1273–1285. doi:10.1068/a42282

Smith, A., Voß, J-P. and Grin, J. 2010. Innovation Studies and Sustainability Transitions: The Allure of the Multi-Level Perspective and Its Challenges. *Research Policy, Special Section on Innovation and Sustainability Transitions* 39, 435–448. doi:10.1016/j.respol.2010.01.023

Star, S.L. 1999. The Ethnography of Infrastructure. *American Behavioral Scientist* 43, 377–391. doi:10.1177/00027649921955326

Swyngedouw, E., Kaika, M. and Castro, E. 2002. Urban Water: A Political Ecology Perspective. *Built Environment* 28(2), 124–137.

Swyngedouw, E. 2004. *Social Power and the Urbanization of Water: Flows of Power*. Oxford University Press, Oxford.

Swyngedouw, E. 2009. The Political Economy and Political Ecology of the Hydro-Social Cycle. *Journal of Contemporary Water Research & Education* 142, 56–60. doi:10.1111/j.1936–1704X.2009.00054.x

Swyngedouw, E. 2015. *Liquid Power: Contested Hydro-Modernities in Twentieth-Century Spain*. MIT Press, Cambridge, MA.

United Nations. 2015. *Sustainable Development Goals* [WWW Document]. Sustainable Development Knowledge Platform, Department of Economic and Social Affairs. https://sustainabledevelopment.un.org/?menu=1300.

United Nations General Assembly. 2000. *United Nations Millennium Declaration* (Resolution adopted by the General Assembly No. A/RES/55/2).

United Nations Sustainable Development. 1992. *Agenda 21*, in: United Nations Conference on Environment and Development. Presented at the United Nations Conference on Sustainable Development, Rio De Janeiro, Brazil.

Warren, K.J. 1990. The Power and the Promise of Ecological Feminism. *Environmental Ethics* 12, 125–146.

Warren, K.J. 1998. Introduction: Ecofeminism, in: Zimmerman, M., Baird Callicott, J, Sessions, G., Warren, K. and Clarke, J. (Eds.), *Environmental Philosophy*. Prentice Hall, Upper Saddle River, pp. 263–276.

WCED. 1987. Towards Sustainable Development, in: World Commission on Environment and Development (Ed.), *Our Common Future*. Oxford University Press, Oxford and New York, pp. 43–65.

5 Demand

Introduction

In cities with growing populations and finite water resources, reducing per capita consumption of water is an obvious strategy for making existing water resources serve more people. If each person uses less, there will be more to go around. Reducing per capita and total demand for water also reduces energy used in treating and pumping water and wastewater, and can delay or offset the need for capital investment in new water supply. Reducing demand for water may be a short-term 'stop-gap' measure during drought and as new supplies are developed, or it can be part of a broader transformation of lifestyles, culture, technologies and infrastructures in cities (Brooks et al., 2009).

Understanding and managing water demand requires fundamentally different approaches to water infrastructure than long-established decision-making processes based on expansion of supply. Water efficiency and demand management campaigns are less certain and reliable in their outcomes than expanding supply because they depend on the performance of fittings and appliances distributed across the city in homes and businesses, and water use is a behavioural and cultural as well as technical phenomenon.

Water wastage was a concern for some of the earliest engineers and planners of urban water infrastructure. A parliamentary debate in England in the 1870s reflected engineers' concerns that moving from an intermittent to a continuous water supply system, particularly without metering, would lead to wastage and inefficiencies (Hillier, 2011). Cities with limited water infrastructure and resources in the Global South today have intermittent water supply, in part to ration limited resources (Andey and Kelkar, 2009). In nineteenth-century London concerns about water wastage were overruled by the benefits of continuous water supply for public health, because an endless, affordable, universal supply of clean water was thought to be essential for improving health. The commitment to public health remains a core principle of water supply, but the inefficiencies and wastage predicted in the nineteenth century have taken on a new urgency as urban water systems reach hydrological, engineering and economic limits.

Water infrastructure in cities in the Global North enables water to be used in homes, public spaces and businesses as if it were an endless resource (Sofoulis, 2005). Metering and pricing provide information to water bill payers that water is a finite commodity, but for most water users, water continues to run as long as the tap is open. As an indicator of the success of modern water infrastructure but a challenge for reducing demand, the lived experience of people in modern cities in the Global North is that taps never run dry. This may be tempered by personal experience of water shortage in other places, stories of water scarcity and bad plumbing from older generations, or care for the local environment, but the message baked into modern water infrastructure is that water is endless (Allon and Sofoulis, 2006).

Patterns of water use are the outcome of interactions between technology, infrastructure, water, daily habits and social and cultural expectations and norms. Strategies for reducing water demand are roughly divided into those that focus on improving water efficiency and those that focus on reducing water use (Russell and Fielding, 2010). Water efficiency measures aim to maintain modern lifestyles and experiences of water, but with a lower resource demand. Strategies for reducing water use address broader cultural and social assumptions and water-using practices, and challenge taken for granted modern expectations for how water is used in cities.

In cities where people live without access to water infrastructure, lack of consumption rather than overconsumption of water remains the key concern. In rapidly urbanising cities of the Global South, infrastructure providers have struggled to keep up with expanding demand. Meeting basic demand for urbanising populations is a significant challenge for engineering, planning, governance and finance. Provision to slums and informal settlements can be particularly difficult given disputes about land ownership, high density of housing, unplanned development, difficulty in raising investment and billing customers, and poor governance (Parikh et al., 2013; Satterthwaite et al., 2005).

It may seem trite to discuss the lack of access to basic water services alongside the challenge of managing demand arising from profligate consumption in the Global North. However, the Global South is not immune to water wastage and hyper-consumption. Water consumption in wealthy households and neighbourhoods may be high, even compared to consumption in the Global North, while people living in poorer areas have difficulty meeting their basic needs (Anand, 2011; Castro, 2004). Informal water networks and poor maintenance can lead to high levels of leakage. Water conservation practices and demand management strategies are therefore of global relevance, particularly in addressing the challenge of serving growing populations within water resource constraints.

Demand for water and the capacity of infrastructure systems to meet it varies widely in cities across the world (Figure 5.1). Many cities have seen

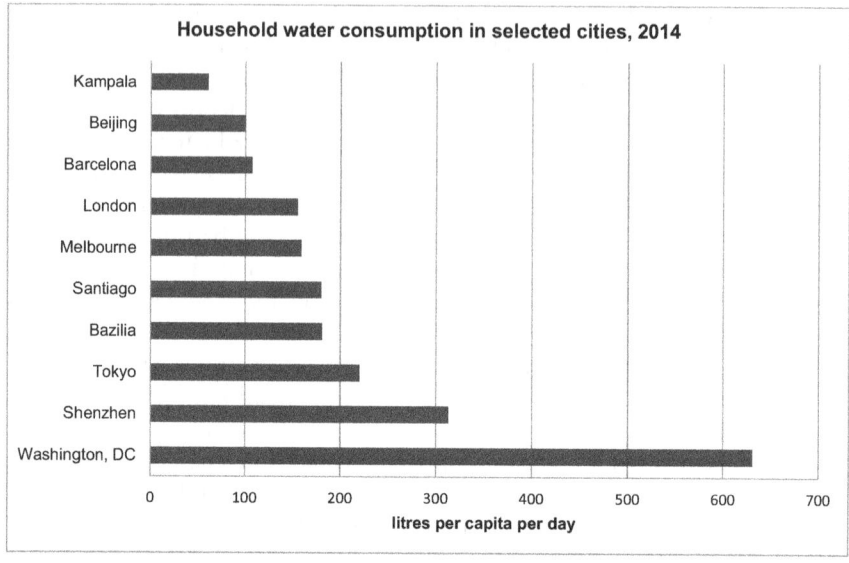

Figure 5.1 Water consumption in selected cities
(Source: IWA, 2017)

reduced per capita consumption in recent years in response to droughts and demand management campaigns, including nearly 60% reduction in per capita consumption in Melbourne and 50% reduction in New York City since the early 1990s (Environment and Natural Resources Committee, 2009; IWA, 2017; NYC Environmental Protection, 2015). The recent history of water consumption in Beijing demonstrates both growth as a result of development and decline as a result of drought and water shortages, increasing from 138 litres per capita per day (lcd) to 238 lcd between 1978 and 1998, and declining to 100 lcd in 2014 (IWA, 2017; Zhang and Brown, 2005).

This chapter focusses on domestic water demand, as this is the most significant element of water consumption in most cities and is therefore of greatest significance in achieving urban sustainability. It describes methods for forecasting water demand, including the key factors that shape demand in the Global North. It then presents different approaches to reducing demand, along a continuum from technical water efficiency to understanding the meaning of water in everyday life. This chapter considers water demand from the perspective of sustainable development, ecological modernisation, socio-technical systems, political ecology and radical ecology, before discussing the role of water demand in sustainable cities.

Forecasting demand

Forecasting demand for water has long been a major task for engineers, planners and water resource managers. Engineers and conservationists have expressed concern about wastage of water and the impact of pollution and abstraction on rivers, streams and wetlands ever since the inception of modern infrastructure, but concerns have been increasing since the 1960s (ICE, 1963; IWPC, 1967). The underlying assumption of mainstream urban water planning has tended to be that water infrastructure would expand to meet demand following a 'predict and provide' philosophy (Butler and Memon, 2006). More recently, the increasing cost of expanding infrastructure to source more water through long-distance transfers, new dams, desalination and potable reuse has helped shift attention towards managing demand. Reducing per capita demand for water in most cases is a low-cost means to ensure that existing resources can meet the needs of a growing population, within local environmental limits.

Forecasting demand for water is a crucial element of water resources and infrastructure planning. Investment cycles in infrastructure require water managers to plan for up to 40 years in advance, amidst uncertainty about populations, politics, the economy and climate (McDonald et al., 2011). On a much shorter timeframe, utility operators need to be sure to have the capacity to deliver water to households and businesses on a daily basis. It is therefore important to have enough water to supply to users and sufficient infrastructure to deliver it. Average long-term trends in total demand drive decisions about water resources, while peak demand drives decisions about the scale, capacity and operation of distribution networks.

Water utilities divide demand into different components. Revenue water is paid for by users, in the household, commercial or industrial sectors, both with and without metering (Farley and Trow, 2003). The remaining components of demand are referred to as non-revenue water and consist of operational water, unbilled use, illegal use and leakage (Frauendorfer and Liemberger, 2010; van den Berg, 2015). Operational water is used by the utility itself in treatment works, pumping stations and their own offices and amenities. Unbilled use includes firefighting and other specific known authorised uses. Illegal use includes unauthorised connections or meter tampering. Leakage as a component of demand refers to losses in the supply and distribution network, though leakage can be significant in homes and businesses. Utility leakage can occur in the distribution network, at storage tanks and up to the point of the customer's connection.

Demand for water follows diurnal and seasonal patterns (McDonald et al., 2011; Parker and Wilby, 2013). Demand is typically lowest at nighttime, and water meters showing high night-time consumption are used to indicate leakage. Demand peaks in the morning as people awake and prepare for work, school and other daily activities, and in the evening as they prepare the evening meal, complete chores and get ready for bed. The daily patterns of water use therefore reflect the rhythms of urban life. Sir Joseph Bazalgette, chief engineer of London's sewers, wrote in 1858 that the peak flows of wastewater in the working-class areas in the east of London occurred around six a.m., and at around nine a.m. in the more middle-class areas in the west of the city (Bazalgette, 1865). Water demand tends to be higher in drier summer months due to increased outdoor use, but it can also be higher in winter due to leaks caused by freezing pipes.

Forecasts of water demand that traditionally depend on demographic and economic forecasts are increasingly informed by detailed knowledge about how water is used in homes and the social, demographic and psychological factors that influence per capita water use (McDonald et al., 2011). Water demand has been correlated to social, demographic and technical factors, but studies of water consumption have shown contradictory results.

In cities in the Global North the one factor that most strongly correlates with per capita consumption is the number of people in the household (Aitken et al., 1994; Willis et al., 2013). While more people in the household intuitively increase the total consumption, per capita consumption is inversely correlated to household size. Single-person households consume more water per person than bigger households. This is consistent with energy consumption trends, though the relationship is less obvious for water than energy. In cities where most energy consumption is associated with heating or air conditioning, the more people sharing spaces such as living rooms and kitchens, the lower the per capita consumption. For water, per capita efficiencies in bigger households may relate to shared garden space, having fuller loads for dishwashing and laundry, and pressure on time in shared bathrooms. Household dynamics have also been shown to influence water consumption, with households that share a culture of water saving using less water (Russell and Fielding, 2010).

While per capita water consumption decreases with the number of people in a household, it increases with the size of the house (Fielding et al., 2012; Inman and Jeffrey, 2006). Water use also tends to increase with income. Bigger houses may have more bathrooms, reducing pressure on shared bathrooms. Households with higher income are able to afford more water-consuming fittings and appliances and to be less constrained by energy and water prices when heating water and running dishwashers, washing machines, showers and baths.

Water demand is also shaped by the climate and built environment (Domene and Saurí, 2006; Saurí, 2013). Residents in hotter, drier climates have higher per capita consumption, and these may also be the cities with the greatest constraints on water resources. People living in apartments, flats or semi-detached houses consume less water per capita than those living in detached houses, particularly with large gardens. Outdoor use in large detached houses is associated with gardening, lawn irrigation, outdoor cleaning, swimming pools and car washing. Urban planning and design therefore influence water demand in similar ways to demand for other resources, which increase in lower-density, car-dependent, sprawling suburbs. This is evident in Barcelona, where the outer suburban regions have more than double the average per capita water consumption than in dense inner city areas (Domene and Saurí, 2006; March et al., 2013; Ostos and Tello, 2014). However, the influence of urban form can be difficult to separate from economic and lifestyle factors that are also associated with different parts of the city and types of development.

Apart from income and household occupancy, other demographic factors are less well correlated with water consumption (Table 5.1). In some studies, older people have been shown to consume more water, while in others they conserve more water than young people (Inman and Jeffrey, 2006). For

Table 5.1 Factors shaping water demand

Factor	Influence on water consumption (per capita)	Reference
Household size	People living in smaller households consumer more water than in bigger households.	(Butler and Memon, 2006)
Income	People living in households with higher income consume more water than those with lower income.	(Fielding et al., 2012)
Age	Mixed. Teenagers are typically high water consumers. Retirees may consume more water at home than people employed outside the home.	(Russell and Fielding, 2010)
House size	People living in large detached houses consume more water than those in smaller houses and apartments.	(Saurí, 2013)
Garden	Garden watering increases water consumption.	(Syme et al., 2004)
Car ownership	Car ownership increases water consumption.	(Domene and Saurí, 2006)
Laundry	Total water consumption is strongly correlated with number of loads of washing per week. Households with more efficient washing machines consume more water overall.	(Aitken et al., 1994) (Fielding et al., 2012)
Climate	Water consumption is higher in hotter, drier climates. Climate change may lead to increased demand in places forecast to experience warmer, drier summers and longer periods without rainfall.	(Domene and Saurí, 2006) (Arnell, 2004)

domestic use 'life stage' may be a better predictor of water use, with high water consumption associated with people spending more time at home after retirement, and with families with teenagers (Fielding et al., 2012).

Water planners use micro-component analysis, also called end-use studies, of water demand to provide breakdowns of how water is used by people in their homes (McDonald et al., 2011; Willis et al., 2013). These methods are based on forecasts of ownership of water-using appliances, such as washing machines and toilets, the typical volume of water per usage and the average frequency of use. This kind of analysis shows how average water use varies in different cities and can be aggregated to forecast future demand. Micro-component approaches to water demand analysis emphasise the importance of water-using appliances and fittings in determining demand, showing how much water is used in toilets, taps, washing machines, showers, baths, outdoors and 'other' unknown uses (see Figure 5.2).

Micro-simulation techniques for forecasting water demand use more detailed statistical methods to integrate trends in demographics and population dynamics and water use (McDonald et al., 2011). Census data and demographic and economic forecasts are used to generate a 'synthetic micropopulation' which is then modelled based on water-using attributes

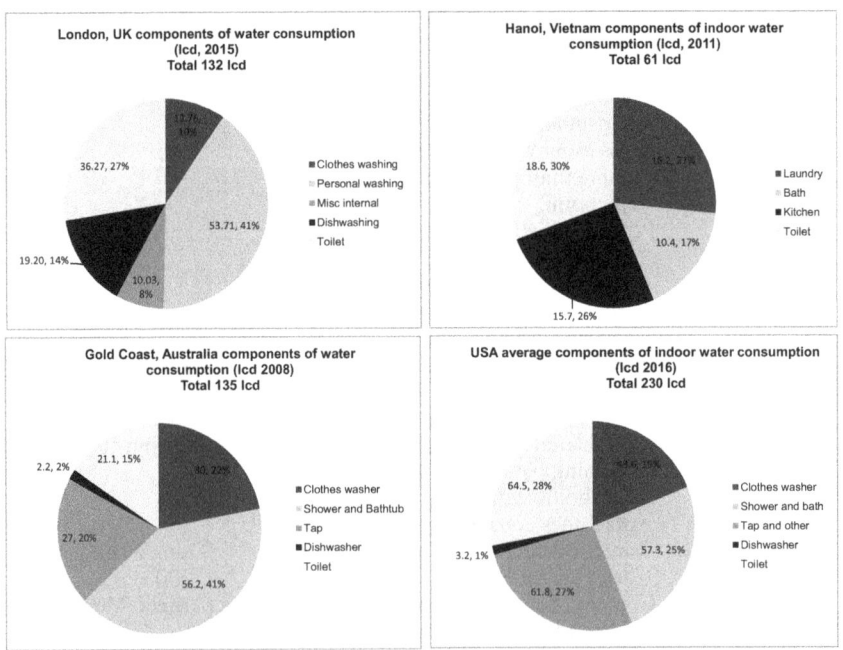

Figure 5.2 International comparison of micro-component water demand

(Sources: DeOreo et al., 2016; Otaki et al., 2008; Thames Water, 2015; Willis et al., 2013)

of particular categories and sizes of households. These simulations can be scaled up to provide a probabilistic forecast of total demand for the future population of the city.

Demand management

Demand management is a central element of urban sustainability. Transport, energy and waste infrastructure face similar challenges to water in reducing or shifting per capita use, consumption and pollution. Technology has a clear role both in creating demand for resources and in improving efficiency of use, but how technologies and infrastructure are used is determined by individual, social and cultural factors.

The 'soft path' for energy management was proposed by Amory Lovins in the 1970s to promote energy efficiency, renewable energy, diversification, decentralisation and demand reduction (Holtz and Brooks, 2009). The soft path for water follows similar principles. Starting from an assumption of 'no new water', water needs must be met by improving efficiency, recycling, rainwater harvesting and changing cultures and lifestyles to reduce water demand. The soft path is in contrast to the hard path of infrastructure construction, particularly dams and desalination plants, and requires greater attention to the needs of water users, stakeholders and the environment than expanding supply (Gleick, 2003).

Demand management is more commonly positioned as one-half of the 'twin-track' approach to increase supply as well as decrease per capita demand (Parker and Wilby, 2013). Whilst 'soft path' approaches look towards a more fundamental reorganisation of infrastructure to reduce absolute demand on resources, within water utilities demand management does not usually replace the need for new supply-side infrastructure, though it may delay new investment and be particularly important as a drought management measure.

Demand management is often promoted as a lower-cost solution to water supply deficits compared to capital investment in supply-side options (Howe and White, 1999). Making the case for demand management as part of an overall strategy for water resources management is dependent on the economic methods used to support decision-making. Least cost planning has been used in Australia and the UK to demonstrate the economic benefits of reducing demand compared to new supply options (Fane et al., 2006; Turner and Jeffrey, 2015). This requires economic analysis to include full costs, including capital, operations and maintenance, and to have data of sufficient robustness to be able to forecast savings from demand management measures. Demand management in the past has been hampered by economic models that place higher value on returns on capital investment rather than conservation measures that may increase operational costs in order to deliver water savings that delay or obviate the need for new capital assets.

Water use restrictions

Reducing demand for water is most pressing during drought events. Water infrastructure systems are generally designed so that restrictions on use are only required every 20–25 years. Drought management plans usually include a series of measures to be implemented as the severity of drought increases, starting from voluntary conservation and education campaigns and moving to progressively stricter limits on non-essential water use. Restrictions on domestic water use generally only apply to outdoor use, particularly bans on sprinklers or garden hoses for all or part of the time, or a complete ban on outdoor water use. Water conservation during drought can lead to longer-term changes in water use patterns. It also demonstrates resilience and adaptability of urban water users to contribute to collective responses to water scarcity.

Experience from droughts in Australia, Spain and England have shown generally high compliance with drought orders to restrict use, even with low levels of enforcement. In Melbourne per capita consumption was more than 400 lcd before the 1997–2009 drought. Through a combination of drought measures and water conservation campaigns, per capita consumption reduced to 276 lcd towards the end of the drought in 2008 and to 159 lcd by 2014 (Environment and Natural Resources Committee, 2009; IWA, 2017). During the 2007–2008 drought in Barcelona water consumption decreased by 21% in one year (March et al., 2013), with some parts of the city consuming just 100 litres per person per day. In London in 2006 average summer consumption fell by more than 5%, with peak consumption reduced by up to 10% (Aylard, 2007).

Theories of water demand

The problem of how to analyse and reduce demand for water in cities has been addressed by different disciplines and professions, each with different formulations of the problem and methods for achieving a solution (Table 5.2). These different approaches can sometimes appear to be in conflict with one another, yet a complementary approach may enable more successful policy and implementation. Hoolohan and Browne (2016) have identified four different categories of collective drivers of water demand: service provision, decision-making, social norms and networks, and sociotechnical practices. Analytical approaches range from the technical to the behavioural and cultural, with most theory and practice recognising the interplay between social and technical factors. The focus of theories and techniques for understanding and reducing demand can be placed on a spectrum moving from efficiency to meaning (Table 5.2).

Leak reduction

Reducing leaks in water infrastructure, plumbing and appliances is a basic first step for reducing demand on water resources. Centralised distribution

Table 5.2 Theories and methods for reducing water demand

Problem definition	Solution orientation	Methods	Dominant discipline
Efficiency	Technology	Leak reduction	Engineering
		Building codes	Planning
		Appliances and fittings	Product design
Behaviour	Individual	Metering	Economics
		Tariffs	Economics
		Labelling	Marketing
		Behaviour change	Psychology
Meaning	Collective	Practices	Sociology
		Cultures	Anthropology

networks are prone to leaks, particularly as some piped networks may be more than 100 years old. Water leakage can occur through corrosion and cracking of pipes, which may be exacerbated by ground conditions, pipe materials and local conditions, such as being subject to high loads beneath roads or driveways. Leaks also occur at pipe joints, valves and in reservoirs and storage tanks (Puust et al., 2010). Leakage is directly related to water pressure, with higher rates of leakage and likelihood of burst pipes when water pressure is higher.

Non-revenue losses and leakage may be the result of technical and managerial factors within the control of the utility, such as pressure, pipe materials and spending on maintenance, but factors such as geography and energy costs are also significant drivers (van den Berg, 2015). In London reducing network pressure has been used as a means of reducing leakage rates, alongside an extensive campaign to replace water mains that were constructed in the nineteenth century. Rather than aiming for zero water leakage, water utilities working under a cost-benefit approach aim to fit and replace pipes until an 'economic level of leakage' is reached – the level beyond which it is more expensive to fix the leaks than to supply extra water. Water leakage is therefore likely to be higher in countries with lower costs of water treatment and pumping, particularly those with low energy costs.

Water leaks outside the distribution network also contribute to water losses and may be more difficult to detect and fix. Leaking pipes, taps, toilets and other appliances rely on building owners noticing leaks and taking action to fix them. If water leaks are causing damage to property or inconvenience to residents, this may prompt a response, but if leaks are undetected or of little consequence then they may remain unattended. Detecting leaks is particularly difficult in properties which are unmetered, including flats, offices and shops which share water meters.

Rates of leakage vary considerably around the world (van den Berg, 2015). Cities with very low levels of leakage include Sydney with around 6% non-revenue water, Singapore with less than 5%, and Amsterdam with less than 3% (Brears, 2017; IBNET, 2017; van den Berg, 2015). In London in 2011 25% of water supplied was lost to leaks, in Rome in 2011 non-revenue water was 37%, and in 2008 non-revenue water in Santiago de Chile was 28% (IBNET, 2017; SWAN Research, 2011; Thames Water, 2015). In 2006 a World Bank study found that non-revenue water was 15% of supply on average in developed countries and 35% of supply in developing countries, with 17% of utilities studied losing more than 50% of supply to non-revenue water (Kingdom et al., 2010).

Building codes

Water efficiency can be promoted through building codes or standards for new buildings and major renovations. Plumbing standards that have previously been based on public health concerns have been adapted to incorporate water efficiency measures. Specifications apply to individual devices or fittings, or overall calculations of building water consumption, which allows designers flexibility in meeting standards for consumption.

Green building certification schemes such as LEED and BREAMM usually include a component related to water efficiency, reuse and rainwater harvesting (Kibert, 2016). The regional variability in water resources presents a challenge for international certification schemes, and LEED has adapted criteria for water for different parts of the world. This is particularly important in schemes that reward reuse or rainwater harvesting systems that may require energy to operate, thereby undermining energy efficiency objectives for the building. Performance-based certification based on micro-component demand forecasting enables flexibility in design, but assumes standardised behaviour and water consumption associated with specific fittings.

Appliances and fittings

Improving the efficiency of household appliances and fittings is important in water demand management, but the presence of water-efficient devices does not always mean lower water consumption (Fielding et al., 2012; Omambala, 2011). Education campaigns may be accompanied by provision of small water-saving devices for users to install in their homes. These include cistern displacement devices, which can be placed in toilet cisterns to reduce the volume of water used for each flush. Low-flow showerheads are also a common device provided to households for free or at a subsidised price.

More extensive demand management campaigns involve replacing existing fittings with more water-efficient devices, such as the toilet replacement programme in New York which installed more than one million water-efficient toilets in three years during in the 1990s as part of a demand management programme that achieved 50% reduction in per capita demand over two decades (Box 5.1). Households can be encouraged or subsidised to replace existing washing machines and dishwashers with more water-efficient models. Improvements in efficiency of toilets and appliances can

---

Box 5.1   New York City water demand management

After more than a century of expansion of supply-side infrastructure, by the 1980s demand for water in New York City exceeded what could be sustainably supplied from the surrounding Catskills, Delaware and Croton catchments (Gandy, 2003; Soll, 2013). The volume of wastewater produced by the city was higher than the capacity of treatment plants, causing untreated wastewater to be discharged into the environment, in breach of state and federal regulations (Soll, 2013). In 1984 water consumption peaked at 947 lcd, and in 1986, following two droughts in five years, the city began a concerted campaign to reduce demand. In 1991 a programme of universal water metering commenced, combined with leak detection, education and water efficiency campaigns targeted at landlords as well as water users (EPA, 2002). A high-profile element of the campaign was a toilet replacement programme, whereby the city provided a rebate of $240 for the replacement of the first toilet in an apartment and $150 for any subsequent replacement (Soll, 2013). Water efficiency measures minimised the impact of water metering on the profitability of landlords and water charges for low-income users, and reduced consumption by nearly 25% by 2003. The reduced demand for water and reduced volume of wastewater led to the cancellation of construction of new water and wastewater treatment plants, representing savings of at least $4 billion. Since then per capita water demand has continued to decline to 476 lcd in 2014. This is still high compared to cities in other countries, but it shows a reduction of 50% over 30 years. Total demand has decreased in New York City since the 1980s by around 30% while population has risen by 19% (NYC Environmental Protection, 2015).

---

contribute to reductions in per capita demand over time as they are replaced as part of their normal lifecycle. Regulations have reduced the volume of toilet flushing since the 1970s, with older toilet flush volumes of up to 20 litres reducing over time to average flushes of 3.5 or 4.5 litres for modern dual-flush toilets (Grant and Moodie, 2002; Keating and Styles, 2004; Schlunke et al., 2008). Washing machines and dishwashers have also consistently improved in water efficiency in recent decades. Improving the water efficiency of appliances must not be at the expense of reduced performance, or there is a risk that overall water consumption will remain high. For instance, low-flush toilets that do not clear the toilet bowl are likely to be flushed twice, and washing machines that do not rinse clothes properly may result in rinse cycles being run again.

Design and installation of water-efficient devices should also consider potential rebound effects. If people are aware that their appliances are water efficient, they may use them more often, negating improvements in efficiency. For instance, if users know that their toilet has a low flush volume, they may flush it unnecessarily to dispose of household waste; they may be less reluctant to wash relatively clean clothes in a water-efficient machine; and they may stand under a low-flow shower for longer.

In a comprehensive study of water use in South East Queensland in 2008 Fielding et al. (2012) confirmed that households with water-efficient devices on the whole are correlated with lower household water consumption, except for washing machines. In that study households with a water-efficient washing machine consumed more water than those with normal washing machines. The authors speculate that this could be due to households with high laundry demands investing in a more efficient machine, or the rebound effect, whereby people use the washing machine more frequently than they otherwise would because they believe that they are saving water by using an efficient machine.

Metering

Water metering is strongly associated with a market-based approach to water efficiency and demand management as it is a prerequisite for providing information about water use to individual households and for using price as a mechanism for reducing water consumption. Water metering constructs water as a commodity, paid for by consumers according to the volume they use. It is also a source of information about water use, enabling comparisons with average or comparable consumption and with historical water use within a household.

Water metering has been associated with reductions in demand, at least in the short term. Inman and Jeffrey (2006) reviewed the impact of metering on consumption in various international studies. Water metering was shown to reduce demand by 10–15% in the UK. Another UK study in 1996 showed that the impact of water metering was strongest in middle-income households consuming 250–400 lcd, with a reduction of 20%, while high-income, high-water-use households reduced consumption by 11%. Water metering had no impact on low-income households consuming less than 250 lcd. In Baltimore in the US, metering was associated with a 56% reduction in outdoor use but no reduction in indoor use. Other studies in the US showed reductions in consumption of between 9–20% associated with metering (Inman and Jeffrey, 2006).

Water metering therefore has a range of impacts on water-using behaviour and per capita water use. The claims for water metering as a demand management measure need to be considered in the context of demographics, pricing, income, behaviour change and housing stock. In the UK houses with meters tend to be newer, and where metering is at the discretion of the householder, meters tend to be installed by people who are already using less water than average or are inclined to change their behaviour to save money. Thus the early experience of high impact of metering may be overstated. It is now recognised that metering must be accompanied by strong customer engagement and water efficiency campaigns in order to achieve reductions in demand.

Smart meters provide utilities with information about water use and demand, which can give detailed and targeted feedback to water users and can be used for variable tariffs (Boyle et al., 2013). Trials of water and energy smart meters show an initial reduction in demand, which may return to normal levels over time. Metering information on its own is unlikely to lead to sustained reductions in water usage, but the data provided by smart meters can be linked to targeted behaviour-change campaigns to provide more customised and timely feedback to water users. As in other sectors, ownership and sharing of smart metering data raises ethical and practical concerns about privacy and security. Detailed information about water use can reveal much about the composition of a household and their daily activities, which they may not wish to share and may leave them vulnerable to criminal activity.

Tariffs

In everyday conversations about water wastage one common complaint is that 'water is too cheap'. For those who believe in the power of the market to drive behaviour, the primary reason for water wastage is that water isn't priced according to its 'true value'. Water pricing can therefore be seen as a straightforward mechanism for managing demand. However, the impact of pricing on demand is more complicated than simplistic economics would presume, and it can have highly uneven consequences for people in different income groups.

Water pricing is a significant concern for equity of access, with highly variable arrangements around the world (Inman and Jeffrey, 2006). In England, water prices are regulated by an independent authority based on water company business plans, which outline capital investments and operating costs. In Ireland, water charges were the source of much controversy following the financial crisis of 2009, and water remains paid for out of general taxation. In South Africa, low-income households are provided with 'free basic water' of 20 lcd, with charges then rising with consumption (Muller, 2008). Cross subsidy of water charges is one mechanism for water utilities to raise funds for investment, by charging higher rates for high water users to enable delivery of basic services at low cost to poorer households.

Variable tariffs have been used as a water demand management method (Herrington, 2007). Rising block tariffs for instance have increasing charges per unit of water as the level of consumption increases. This still allows for basic needs to be met at a relatively low cost, but it provides penalties for increasingly profligate use. Water charges might also be varied seasonally to allow utilities to charge more during times of water shortage.

The use of tariffs as a demand management technique is constrained by the relative inelasticity of water demand (Inman and Jeffrey, 2006). Lower-income and larger households, particularly those in flats or smaller houses, have little capacity to reduce water use, and rising tariffs will simply mean an increase in spending on water. For higher-income households, even those with high discretionary use of water such as watering large gardens and maintaining swimming pools, water charges are a relatively insignificant element of their household spending, and increases in tariffs will have relatively little impact. Water metering and tariffs can discourage profligacy in middle-income households, achieving demand management objectives, but water pricing itself is a relatively blunt measure for reducing demand.

Labelling

Providing water efficiency information by labelling devices is important to allow consumers to take account of water efficiency in their purchasing, but this will only be one element of their purchasing decision (Chong et al., 2008). Water efficiency labelling schemes in Singapore and Australia provide ratings based on water use, similar to energy star rating schemes that are common for electrical appliances. Water efficiency labelling is most effective when it is compulsory for all retail products. Voluntary schemes such as the Waterwise label in the UK can be beneficial in highlighting the most efficient devices to consumers, but without comparison across all products on the market consumers are not provided with complete information to inform purchasing decisions. Water efficiency labelling drives manufacturers to innovate to avoid low ratings and to achieve market leadership. As with energy efficiency labelling, water labelling can drive the most wasteful products out of the market. As the overall efficiency of devices increases, labelling schemes need to be revised to allow for continued product differentiation and drive continued innovation. The most efficient devices on the market in the 1990s may be the least efficient 30 years later, and labelling schemes therefore need to be revised periodically.

Behaviour change

Behaviour change theories and models are devised by psychologists and behavioural economists. They have been developed across a number of sectors including health, transport, energy, and waste management, as well as in water conservation. Public health practitioners are interested in reducing unhealthy behaviours such as smoking and excessive alcohol consumption and promoting healthy behaviours such as exercise and a balanced diet. Programmes to change travel behaviour away from private vehicles towards greater uptake of public transport, walking and cycling also draw on psychological and social science theories. Similarly, behaviour change aims to reduce energy-consuming behaviours and to encourage positive waste management behaviours such as recycling.

Behaviour change in water demand management focusses on inefficient or wasteful everyday water use, such as leaving taps running, using dishwashers and washing machines on half loads, and taking lengthy showers and frequent baths. Behaviour change models are also used to influence longer-term decision-making and trends such as garden design and plantings, and purchasing decisions for appliances or major home renovations.

Behaviour change programmes focus on individual attitudes, beliefs, perceptions, motivations and intentions as the basis for individual behaviours. The purpose is to understand the causes of behaviour in order to identify opportunities to influence change. For instance, Ajzen's Theory of Planned Behaviour identifies attitudes, subjective norms and perceived behavioural control as the three key factors influencing individuals' intentions, which then contribute to the behaviour (Ajzen, 2011, 1985). More elaborate models, such as the Behaviour Change Wheel developed by Susan Michie and colleagues, address the social, policy, economic and environmental contexts that enable and constrain desired behaviours (Michie et al., 2011; Michie and Johnston, 2012). In a review of environmental behaviour change research Steg and Vlek (2009) distinguish between motivational factors, contextual factors and habitual behaviour as contributors to behaviour. They divide interventions into information strategies which seek to influence the individual's knowledge, attitudes, motivations and beliefs, and structural strategies that seek to change the context in which behaviours take place.

Contextual factors driving water use behaviour include pricing, metering, household size, appliance efficiency and others addressed earlier in this chapter (Russell and Fielding, 2010). Information strategies include public awareness campaigns, education in schools and home visits. Programmes based on social marketing and social norms utilise social networks to share information about water conservation behaviours and to shift beliefs about shared values and 'normal' behaviour. Attitudes and values towards the environment and water have been shown to influence water conservation behaviour (Russell and Fielding, 2010). Information, values and social norms can contribute to changing both behaviour and cultural trends, such as choices to xeriscape gardens through the selection of drought-tolerant or low-water-demanding plants and landscaping (Syme et al., 2004).

Practices

The behaviour change approach has been criticised for failing to address the importance of relationships between technology, infrastructure, culture and consumption. Practice theory provides an alternative analysis of resource consumption, focussing on everyday life rather than individual behaviour. While a behavioural theory may conceive of showering as a water-using behaviour, practice theory analyses it as a personal cleanliness and relaxation practice (Shove, 2010, 2004). How often and how long people shower are more likely to be shaped by social expectations for cleanliness, a daily routine associated with waking up or bedtime, and the pleasurable experience of standing under hot running water than an individual's attitude to water or the environment.

Practice-based approaches to water demand bear similarities to structural approaches to behavioural models. However, they are distinct in focussing on the 'practice' associated with water consumption, such as personal cleanliness, rather than the behaviour, such as length of shower (Shove, 2004). Paying attention to practices rather than behaviours not only demonstrates the complexity of interactions that drive water consumption in households, but it also identifies alternative practices that are less resource intensive. Shove (2004) demonstrated the importance of social and cultural expectations which shape everyday water-using practices and have co-evolved with technologies and infrastructures. Achieving significant, long-term reductions in per capita demand for water requires redesigning water systems to account for the connections between culture, technology, infrastructure and water-using practices (Browne, 2015). This requires reconfiguring infrastructure and household systems to not only conserve water, but also to shift expectations and practices that lead to high water consumption.

Understanding water consumption through the lens of social practices promotes a greater attention to diversity than the standard behavioural models (Shove, 2010). Water-using practices not only change over time but vary within and between communities. Communities and families with experience of drought or water scarcity may have developed alternative practices for cleanliness, gardening or relaxation that are less water intensive and may provide the basis for more widespread change. A practice-based approach also draws attention to the need for technologies and infrastructures to support water conservation, rather than narrow behavioural approaches that place responsibility for resource use on individuals' attitudes and beliefs.

Cultures

Water-using behaviours and practices occur within wider cultures of consumption and everyday life. Cultures can drive social norms that can influence water conservation by valuing the environment and admonishing wastage, but may also drive up consumption through expectations of cleanliness and design fashions in gardening, bathrooms and modern appliances (Allon and Sofoulis, 2006; Askew and McGuirk, 2004).

Water is a powerful symbol in most religious and spiritual traditions. Water conservation campaigns run in partnership with churches, mosques, synagogues and other religious communities draw on scriptural and theological values associated with water specifically and conservation more generally. Within religious groups, regional and ethnic diversity can influence values and practices. For instance, a water and faith campaign instigated by Thames Water in London designed information and behaviour change campaigns that addressed specific needs and values of different ethnic and religious groups. Muslim people of South Asian origin may share water-using practices based on faith with Muslim people of African heritage, but they will also share practices with ethnically South Asian people of Sheik or Hindu faith based on cooking or cleanliness practices.

Cultures are drivers of water-using practices, particularly in relation to gardening and homemaking. Askew and McGuirk (2004) analysed gardening in a new suburban development in Australia as a means for accumulating cultural capital. Choices about garden design and plantings reflect cultural distinction, associated with high status or particular values such as environmental conservation, and cultural conformity, associated with the need to 'fit in' with shared norms. Gardens are places of relaxation, leisure and socialising, and they may be conceived of as an extension of the living area beyond the house. Well-kept lawns and gardens are a source of cultural capital, and garden design and maintenance is an important source of meaning and expression of values and identity. Experiences of drought and water conservation campaigns can drive changes in garden fashions towards lower water consumption, but this may not be universal or permanent. Fashions for water features, tropical plants, swimming pools and manicured lawns are cultural phenomena associated with high water consumption.

Framing water demand

Demand for water is a key driver of urban sustainability. Reducing water wastage and improving efficiency are widely agreed to be necessary to reduce the need to expand supply. However, different theoretical and political framings emphasise different priorities and mechanisms for change. Specific theories of water demand may be mapped onto wider frameworks for urban sustainability, which provide a broader context for the diversity of approaches.

Sustainable development

Through its emphasis on reducing poverty and improving public health, sustainable development prioritises meeting demand for basic water and sanitation for those who currently have no or insufficient access. Sustainable Development Goal 6 has as its first target: 'By 2030, achieve universal and equitable access to safe and affordable drinking water for all'. The SDGs recognise the need for water efficiency with a further target: 'By 2030, substantially increase water-use efficiency across all sectors and ensure sustainable withdrawals and supply of freshwater to address water scarcity and substantially reduce the number of people suffering from water scarcity' (United Nations, 2015). Improving water efficiency is seen as a measure to reduce water scarcity, ensuring resources are available to provide universal access.

Inefficiency and lack of access can both be outcomes of poor infrastructure management and governance (Bakker, 2010). Cities with the lowest levels of access to water are also often associated with high levels of leakage and non-revenue losses (Kingdom et al., 2010). Lack of access in slums can drive water theft, which may be associated with wider criminal networks or may simply be a survival strategy for residents (Anand, 2011). Sustainable development projects and programmes, including those funded by international donors, emphasise the construction of water infrastructure to support economic growth and universal access. Ensuring good governance, including mechanisms for ongoing financing for infrastructure services, has been seen as an important element of infrastructure projects to deliver sustainable water services to meet demand.

The minimum volume of water required to meet basic human needs is contested. The World Health Organisation recommends 20 lcd in emergency situations to meet basic needs for drinking and hygiene. Supporting reasonable livelihoods and an acceptable standard of living is thought to require 80–100 lcd, and up to 135 lcd is required to meet economic development and agricultural needs (Chenoweth, 2008).

Processes of urbanisation are likely to continue to drive demand for water in cities. Sustainable management of resources requires that demand does not exceed the volume of water that can be safely abstracted from the environment. Achieving 80–100 lcd for the poorest residents of a city may require reducing the consumption of wealthier residents. Beijing's history of average consumption rising from 138 lcd to 238 lcd between 1978 and 1998, and declining to 100 lcd in 2014, demonstrates that average water demand can rise and fall significantly within a generation (IWA, 2017; Zhang and Brown, 2005).

Metering, tariffs and a range of programmes can be used to drive efficient use of water and also enable cross subsidy, whereby higher water charges paid by high-volume users are used to finance lower-cost basic access. The South African 'free basic water' policy is based on a right to water for basic needs only (Muller, 2008). Implementation of the policy has been controversial, particularly in the definition of 'basic water', with low-income households, particularly large households, unable to meet needs within the lower limit and unable to pay for the steeply rising block tariff.

In the Global North reducing per capita water demand is a relatively uncontroversial element of sustainable water management. Reducing demand can be seen to meet social, environmental and economic objectives within a framework of sustainable development. However, strategies for implementation and the relative priority of reducing demand compared to increasing supply are less widely agreed upon, as evident in the range of approaches presented in this chapter.

Ecological modernisation

Water demand management based on water-efficient technology, metering and billing is consistent with ecological modernisation theory. Water-efficient technologies promise that current lifestyles can be maintained at the same time as reducing domestic water demand, and behaviour change campaigns focus on education of consumers about the value of water and how they can reduce wastage. Installing water-efficient technologies can be the responsibility of the individual householders, water companies and builders. Water efficiency is governed through building codes and building assessment tools, water company targets set by regulators, and local government planning.

Proponents of water efficiency highlight the need for reform of economic decision-making and accounting. Least cost option planning, lifecycle costing and including capital and operating expenditure on equal terms in business planning and water price setting have contributed to making the economic case for water efficiency within utilities and regulatory authorities. The development of economic instruments and tools to demonstrate the value of water efficiency is consistent with ecological modernisation approaches that emphasise reform of existing markets and institutions rather than radical overall or strict regulation.

The decoupling of economic and population growth from water consumption as seen in New York, Beijing, Melbourne and other cities demonstrates the goals of ecological modernisation theory are feasible. Regulation, restrictions, technological innovation, behaviour change and a range of policy instruments involving private, public and civil society stakeholders contribute to reversing twentieth-century trends of increasing per capita consumption with increasing wealth. Whilst the Ecomodernist Manifesto explicitly focusses on innovation in supply-side technologies such as desalination to overcome water resource constraints, evidence from cities around the world shows the role of demand-side measures in achieving improved living standards within existing hydrological constraints (Asafu-Adjaye et al., 2015).

Modernisation and expansion of inadequate and decrepit water supply networks in cities in the Global South may also deliver improved access to water while reducing losses due to leakage. The private sector has a role to play in design, construction, delivery and management of water systems, but for ecological modernisation to achieve sustainable and equitable outcomes this must occur within strong governance structures to secure affordable tariffs for the urban poor, high water quality standards, and ongoing funding for maintenance and operations.

Socio-technical systems

Water demand is widely recognised to consist of both social and technical elements. Socio-technical transitions theory aims to explain how water efficiency innovations in both technologies and practices are enabled or constrained by wider institutional and cultural contexts (Geels, 2011). Water-efficient devices such as low-flush toilets or low-water-using washing machines emerge as 'niche' innovations to solve a particular problem, then may be expanded by regulations, labelling and water conservation policies at the 'regime' level, and normalised by acceptance as part of a wider political and cultural acceptance of water efficiency as a core element of modern society. Isolated innovation, changing practices and product development are necessary but not sufficient to achieve transition to a water-efficient society and economy.

A key insight from socio-technical approaches to water demand has been to show the connection between infrastructure and consumption. Supply and demand, and infrastructure and behaviour, are not separate realms to be dealt with in isolation, but are deeply connected. Patterns of demand and water-using practices have co-evolved with the provision of continuous water and sewerage services, and infrastructures have in turn grown and changed to meet demand. Everyday practices of cleanliness have evolved over time as infrastructure and technology have enabled different social and cultural expectations (Shove, 2004).

Shove (2004) uses the example of bathing to make this point. 'Normal' bathing practices in the UK have evolved from a weekly bath and daily sponge-wash in the mid-twentieth century to daily showering. As water supply and heating infrastructure have become more reliable, bathrooms have become more important in homes and new products such as shower gels have come to market. The move from a weekly bath to a daily shower has driven up demand for water in cities, and so efforts to improve water sustainability should be able to account for such complex interactions and transformation that underpin patterns of consumption. Behavioural campaigns to reduce water consumption focus on changing attitudes and behaviours such as reducing time spent in the shower. Practice-based approaches might investigate opportunities to reinvent marginalised cleanliness practices, such as the daily sponge-wash, or to promote less resource-intensive, socially acceptable alternatives for everyday relaxation practices, such as lying in bed rather than standing under the shower as part of a morning routine.

Demand management programmes that frame bathing as an individual behaviour isolated from technology, culture and infrastructure are unlikely to achieve significant long-term transformation of water consumption. The socio-technical approach increases the complexity of understanding how water is used and how water demand relates to water infrastructure and wider cultures. These insights may appear difficult to operationalise within existing policy, planning and engineering frameworks, but they provide the basis for alternative policy and design practice.

Political ecology

The wide global inequality in domestic water consumption can be seen as a proxy for broader global patterns of unequal development (Figures 5.1 and 5.2). Within cities, inequality of access to water and the high absolute and relative price that low-income groups pay for water also demonstrate wider patterns of inequality. Political ecology draws attention to such uneven distribution of costs and benefits of water infrastructure and seeks to identify the underlying political and economic trends that are at its root.

Political ecologists point to neoliberal policies, including privatisation and financialisation of water infrastructure, as key drivers for inequality and environmental harm arising from water infrastructure development and management. Capital accumulation inherently favours building new assets such as dams and desalination plants over demand-side management options (Loftus, 2009). As water infrastructure has become fully or partially privatised it is increasingly attractive to global investment funds as a secure and stable asset and source of reliable return. Governance arrangements that transfer the cost of paying for capital projects to water users through increased bills provide further security on return for investors and private companies, through long-term contracts for supply. These broad policy approaches create a context in which expanding supply is more attractive than reducing demand, which will lead to a reduction in water bills and revenue for water utilities. For utilities aiming to maximise capital value and minimise operating expenses, demand management may not be as attractive as investing in new supply options, despite overall benefits. Demand management reduces large-scale capital investment and distributes spending on new technologies and behaviour change throughout individual households.

Demand management measures based on water metering and tariffs have also been critiqued by political ecologists as unfairly impacting on lower-income households. Water metering as a demand management measure assumes that consumers behave rationally and respond to more information about their water use, including price signals. This is an individualistic, neoliberal model of human behaviour and decision-making, which is uncontroversial within much engineering systems thinking but does not fully account for the complexity and diversity of factors influencing how people use water. For Matthew Gandy (2002) the implementation of universal water metering in New York City in the 1990s was a socially regressive measure, partly in response to the decline in capital investment in water infrastructure during the 1970s and '80s in the early decades of deregulation of global capital markets (Box 5.1). Universal metering and water charges are therefore consistent with neoliberal politics of shifting the cost of infrastructure from the state to individual consumers, with particularly regressive consequences for the urban poor.

Radical ecology

From a radical ecology perspective, water provides the basis for reconsidering human relationships with the natural world. Water provides a direct connection between local landscapes and hydrology and local communities and cultures. Demand for water should therefore adapt to fit within local hydrological limits, providing sufficient water for basic human and ecosystem needs. Changing seasonal patterns of water use and reducing demand during drought are indications of learning to live within local ecological networks and resources.

A radical ecology framing opens up a wider set of possibilities for reducing demand than conventional strategies for efficiency and behaviour change. UK-based artist Chloe Whipple undertook a personal experiment and artistic project in 2015–2016 to live on just 15 litres of water per day for a year, less than the WHO recommended minimum of 20 litres. Whipple's online account of her progress shows motivation to 'demonstrate that society has become disconnected with the water that we consume through living on 10% of a typical UK consumer's water requirements' (Whipple, 2016). This radical transformation of consumption required dramatically altered practices of cleanliness, including personal washing and laundry. Whipple partnered with water engineer Peter Melville-Shreeve to implement simple water-saving and recycling technologies to help her achieve her goal, demonstrating a role for technology in supporting alternative practices and lifestyles. Whipple's intention was to demonstrate not simply the possibility of living in an industrialised society using very little water, but a wider ethic of care for the environment. As she wrote on her blog, 'I want to know that when I leave this place I tried my hardest to live in a way that didn't disrespect and destroy the beauty and abundance of my home'.

Whipple's intention was not to transform society or infrastructure but, as an artist, to demonstrate the possibility of approaching water from a radically different perspective to the dominant culture she lives within. Living successfully on 15 litres per day required strong personal commitment and accountability, as well as support from family, friends and Melville-Shreeve, the engineer. Whipple was not campaigning for everyone in the UK to reduce water consumption by 90%; she was merely demonstrating that it is possible, and that a commitment to low water use leads to a greater sense of connection and care for the local environment, or 'home' as she calls it.

Radical ecology, in emphasising the role of water as a connector between people, cities and landscapes, and in promoting an ethic of care for local places, supports dramatic changes in lifestyles, technologies and systems. Whilst practices such as Chloe's experiment are unlikely to become mainstream in the UK under current conditions, pushing the boundaries of what is acceptable and achievable highlights the possibilities for dramatically changing technology and lifestyles to live within local constraints even under extreme water scarcity.

Demand in sustainable cities

Demand is the foundation of urban water sustainability. For cities in the Global South, this means providing infrastructure to ensure fair access to water to support health and livelihoods. In the Global North, reducing per capita demand is the first priority for sustainable urban water management. Ensuring people have enough water to lead healthy, prosperous lives within the constraints of available water resources is a key challenge for urban engineers and planners. As urban populations grow, this task becomes more complex. Providing sufficient water to support wellbeing in cities without encouraging profligacy and wastage requires collaboration and deliberation, as well as technical innovation.

Water demand is shaped by a range of technical, economic, demographic, social and cultural factors. Understanding and managing water demand is similar to other urban resource efficiency strategies, such as energy, waste and transport. Common approaches to behaviour change, tariffs, metering and technical efficiency can benefit urban sustainability. Water efficiency campaigns have delivered dramatic reductions in per capita consumption, allowing cities to accommodate growing population and to adapt to water scarcity.

Water is unique as it is critical to human health and is mostly constrained by local resource availability. Water and water scarcity also have long-standing cultural associations that are more deeply held than for energy or transport infrastructure. During drought urban citizens have shown capacity to radically reduce consumption and adapt lifestyles and technologies to live within hydrological constraints. This provides opportunities for broader discussion and debate about fundamental reconfiguration about how water is used in cities and its role in everyday life. Engaging citizens in the task of figuring out how to live well in cities within local hydrological limits points to the core of the challenge of urban sustainability, balancing personal responsibility, shared cultures, infrastructures and physical resource constraints.

References

Aitken, C.K., McMahon, T.A., Wearing, A.J. and Finlayson, B.L. 1994. Residential Water Use: Predicting and Reducing Consumption. *Journal of Applied Social Psychology* 24, 136–158. doi:10.1111/j.1559–1816.1994.tb00562.x

Ajzen, I. 1985. From Intentions to Actions: A Theory of Planned Behavior, in: Kuhl, J. and Beckmann, J. (Eds.), *Action Control, SSSP Springer Series in Social Psychology*. Springer, Berlin and Heidelberg, pp. 11–39.

Ajzen, I. 2011. The Theory of Planned Behaviour: Reactions and Reflections. *Psychology & Health* 26, 1113–1127. doi:10.1080/08870446.2011.613995

Allon, F. and Sofoulis, Z. 2006. Everyday Water: Cultures in Transition. *Australian Geographer* 37, 45–55. doi:10.1080/00049180500511962

Anand, N. 2011. PRESSURE: The PoliTechnics of Water Supply in Mumbai. *Cultural Anthropology* 26, 542–564. doi:10.1111/j.1548–1360.2011.01111.x

Andey, S.P. and Kelkar, P.S. 2009. Influence of Intermittent and Continuous Modes of Water Supply on Domestic Water Consumption. *Water Resources Management* 23, 2555–2566. doi:10.1007/s11269-008-9396-8

Arnell, N.W. 2004. Climate Change and Global Water Resources: SRES Emissions and Socio-Economic Scenarios. *Global Environmental Change* 14, 31–52. doi:10.1016/j.gloenvcha.2003.10.006

Asafu-Adjaye, J., Blomqvist, L., Brand, S., Brook, B., Defries, R., Ellis, E., Foreman, C., Keith, D., Lewis, M., Mark, L., Nordhaus, T., Pielke Jr, R., Pritzker, R., Roy, J., Sagoff, M., Shellenberger, M., Stone, R. and Teague, P. 2015. *An Ecomodernist Manifesto*. www.ecomodernism.org

Askew, L.E. and McGuirk, P.M. 2004. Watering the Suburbs: Distinction, Conformity and the Suburban Garden. *Australian Geographer* 35, 17–37. doi:10.1080/0004918024000193702

Aylard, R. 2007. *Delivering Water Efficiency: The Thames Water Experience*. Presented at the Waterwise 2007 Conference: The Economics of Water Efficiency in the Natural and Built Environment, Oxford.

Bakker, K. 2010. *Privatizing Water*. Cornell University Press, Ithaca and London.

Bazalgette, J.W. 1865. On the Main Drainage of London: And the Interception of the Sewage from the River Thames. *Minutes of the Proceedings of the Institution of Civil Engineers* 24, 280–314.

Boyle, T., Giurco, D., Mukheibir, P., Liu, A., Moy, C., White, S. and Stewart, R. 2013. Intelligent Metering for Urban Water: A Review. *Water* 5, 1052–1081. doi:10.3390/w5031052

Brears, R.C. 2017. *Urban Water Security*. John Wiley & Sons, Chichester.

Brooks, D. Brandes, O. and Gurman, S. 2009. *Making the Most of the Water We Have*. Earthscan, London and Washington, DC.

Browne, A.L. 2015. Insights from the Everyday: Implications of Reframing the Governance of Water Supply and Demand from "People" to "Practice." *WIREs Water* 2, 415–424. doi:10.1002/wat2.1084

Butler, D. and Memon, F.A. 2006. Water Consumption Trends and Demand Forecasting Techniques, in: Butler, D. and Memon F. (Eds.), *Water Demand Management*. IWA Publishing, London, pp. 1–26.

Castro, J.E. 2004. Urban Water and the Politics of Citizenship: The Case of the Mexico City Metropolitan Area during the 1980s and 1990s. *Environmental and Planning A* 36, 327–346. doi:10.1068/a35159

Chenoweth, J. 2008. Minimum Water Requirement for Social and Economic Development. *Desalination* 229, 245–256. doi:10.1016/j.desal.2007.09.011

Chong, J., Kazaglis, A. and Giurco, D. 2008. *Cost-Effectiveness Analysis of WELS: The Water Efficiency Labelling and Standards Scheme (Report)*. Institute for Sustainable Futures, University of Technology, Sydney.

DeOreo, W., Mayer, P., Dziegielewski, B. and Keifer, J. 2016. *Residential End Uses of Water, Version 2: Executive Summary*. Water Research Foundation, Denver.

Domene, E. and Saurí, D. 2006. Urbanisation and Water Consumption: Influencing Factors in the Metropolitan Region of Barcelona. *Urban Studies* 43, 1605–1623. doi:10.1080/00420980600749969

Environment and Natural Resources Committee. 2009. *Inquiry into Melbourne's Future Water Supply*. Parliament of Victoria, Melbourne.

EPA. 2002. *Cases in Water Conservation (No. EPA832-B-02–003)*. United States Environmental Protection Agency, Washington, DC.

Fane, S., Turner, A. and Mitchell, C. 2006. The Secret Life of Water Systems: Least Cost Planning Beyond Demand Management, in: Beck, M. and Speers, A. (Eds.), *2nd IWA Leading-Edge Conference on Sustainability in Water Limited Environments*. IWA Publishing, London, pp. 35–41.

Farley, M. and Trow, S. 2003. *Losses in Water Distribution Networks*. IWA Publishing, London.

Fielding, K.S., Russell, S., Spinks, A. and Mankad, A. 2012. Determinants of Household Water Conservation: The Role of Demographic, Infrastructure, Behavior, and Psychosocial Variables. *Water Resources Research* 48, W10510. doi:10.1029/2012WR012398

Frauendorfer, R. and Liemberger, R. 2010. *The Issues and Challenges of Reducing Non-Revenue Water*. Asian Development Bank, Mandaluyong City.

Gandy, M. 2002. *Concrete and Clay*. MIT Press, Cambridge and London.

Gandy, M. 2003. *Concrete and Clay: Reworking Nature in New York City*. MIT Press, Cambridge, MA.

Geels, F.W. 2011. The Multi-Level Perspective on Sustainability Transitions: Responses to Seven Criticisms. *Environmental Innovation and Societal Transitions* 1, 24–40. doi:10.1016/j.eist.2011.02.002

Gleick, P.H. 2003. Global Freshwater Resources: Soft-Path Solutions for the 21st Century. *Science* 302, 1524–1528. doi:10.1126/science.1089967

Grant, N. and Moodie, M. 2002. Focus – Water Closets: Best Practice Since the Water Fittings Regulations 1999. *GreenPro News* 12–14.

Herrington, P. 2007. *Waste Not, Want Not? Water Tarriffs for Sustainability*. A report to WWF-UK. Centre for Sustainable Energy, Bristol.

Hillier, J. 2011. The Rise of Constant Water in Nineteenth-Century London. *The London Journal* 36, 37–53. doi:10.1179/174963211X12924714058689

Holtz, S. and Brooks, D. 2009. In the Beginning: Soft Energy Paths, in: Brooks, D. (Ed.), *Making the Most of the Water We Have: The Soft Path to Water Management*. Earthscan, London and Sterling, VA, pp. 35–48.

Hoolohan, C. and Browne, A.L. 2016. Reframing Water Efficiency: Determining Collective Approaches to Change Water Use in the Home. *British Journal of Environment and Climate Change* 6, 179–191.

Howe, C. and White, S. 1999. Integrated Resource Planning for Water and Wastewater. *Water International* 24, 356–362. doi:10.1080/02508069908692188

IBNET. 2017. *IB-NET Database. International Benchmarking Network for Water and Sanitation Utilities*. www.ib-net.org

ICE. 1963. *Conservation of Water Resources in the United Kingdom: Proceedings of the Symposium Organised by, and Held at, the Institution of Civil Engineers on 30 and 31 October and 1 November 1962*. Institution of Civil Engineers, London.

Inman, D. and Jeffrey, P. 2006. A Review of Residential Water Conservation Tool Performance and Influences on Implementation Effectiveness. *Urban Water Journal* 3, 127–143. doi:10.1080/15730620600961288

IWA. 2017. *International Statistics for Water Services* [WWW Document]. International Water Association. www.waterstatistics.org

IWPC. 1967. *Conservation and Reclamation of Water*. Institute of Water Pollution Control, S.l.

Keating, T. and Styles, M. 2004. *Performance Assessment of Low Volume Flush Toilets*. Southern Water, Worthing.

Kibert, C.J. 2016. *Sustainable Construction: Green Building Design and Delivery*. John Wiley & Sons, Chichester.

Kingdom, B., Limberger, R. and Mari, P. 2010. *The Challenge of Reducing Non-Revenue Water (NRW) in Developing Countries, Water and Sanitation Discussion Paper*. World Bank Group, Washington, DC.

Loftus, A. 2009. Rethinking Political Ecologies of Water. *Third World Quarterly* 30, 953–968. doi:10.1080/01436590902959198

March, H., Domènech, L. and Saurí, D. 2013. Water Conservation Campaigns and Citizen Perceptions: The Drought of 2007–2008 in the Metropolitan Area of Barcelona. *Natural Hazards* 65, 1951–1966. doi:10.1007/s11069-012-0456-2

McDonald, A., Butler, D. and Ridgewell, C. 2011. Water Demand: Estimation, Forecasting and Management, in: Savic, D. and Banyard, J. (Eds.), *Water Distribution Systems*. Thomas Telford Ltd, London, pp. 49–71.

Michie, S. and Johnston, M. 2012. Theories and Techniques of Behaviour Change: Developing a Cumulative Science of Behaviour Change. *Health Psychology Review* 6, 1–6. doi:10.1080/17437199.2012.654964

Michie, S., van Stralen, M.M. and West, R. 2011. The Behaviour Change Wheel: A New Method for Characterising and Designing Behaviour Change Interventions. *Implementation Science* 6, 42. doi:10.1186/1748-5908-6-42

Muller, M. 2008. Free Basic Water – A Sustainable Instrument for a Sustainable Future in South Africa. *Environment and Urbanization* 20, 67–87. doi:10.1177/0956247808089149

NYC Environmental Protection. 2015. *Water Demand Management Report. June 2015 Update*. New York City Environmental Protection, New York City.

Omambala, I. 2011. *Evidence Base for Large-Scale Water Efficiency Phase II Final Report*. Waterwise, London.

Ostos, J.R. and Tello, E. 2014. A Long-Term View of Water Consumption in Barcelona (1860–2011): From Deprivation to Abundance and Eco-Efficiency? *Water International* 39, 587–605. doi:10.1080/02508060.2014.951252

Otaki, Y., Otaki, M., Pengchai, P., Ohta, Y. and Aramaki, T. 2008. Micro-Components Survey of Residential Indoor Water Consumption in Chiang Mai. *Drinking Water Engineering and Science* 1, 17–25. doi:10.5194/dwes-1–17–2008

Parikh, P., Parikh, H. and McRobie, A. 2013. The Role of Infrastructure in Improving Human Settlements. *Proceedings of the Institution of Civil Engineers – Urban Design and Planning* 166, 101–118.

Parker, J.M. and Wilby, R.L. 2013. Quantifying Household Water Demand: A Review of Theory and Practice in the UK. *Water Resource Management* 27, 981–1011. doi:10.1007/s11269-012-0190-2

Puust, R., Kapelan, Z., Savic, D.A. and Koppel, T. 2010. A Review of Methods for Leakage Management in Pipe Networks. *Urban Water Journal* 7, 25–45. doi:10.1080/15730621003610878

Russell, S. and Fielding, K. 2010. Water Demand Management Research: A Psychological Perspective. *Water Resources Research* 46, W05302. doi:10.1029/2009WR008408

Satterthwaite, D., McGranahan, G. and Mitlin, D. 2005. *Community-Driven Development for Water and Sanitation in Urban Areas*. International Institute for Environment and Development, London.

Saurí, D. 2013. Water Conservation: Theory and Evidence in Urban Areas of the Developed World. *Annual Review of Environment and Resources* 38, 227–248. doi:10.1146/annurev-environ-013113–142651

Schlunke, A., Lewis, J. and Fane, S. 2008. *Analysis of Australian Opportunities for More Water Efficient Toilets. Institute for Sustainable Futures*, University of Technology Sydney, Sydney.

Shove, E. 2004. *Comfort, Cleanliness and Convenience: The Social Organization of Normality*. Berg Publishers, Oxford.

Shove, E. 2010. Beyond the ABC: Climate Change Policy and Theories of Social Change. *Environment and Planning A* 42, 1273–1285. doi:10.1068/a42282

Sofoulis, Z. 2005. Big Water, Everyday Water: A Sociotechnical Perspective. Continuum: *Journal of Media & Cultural Studies* 19, 445–463. doi:10.1080/1030 4310500322685

Soll, D. 2013. *Empire of Water: An Environmental and Political History of the New York City Water Supply*. Cornell University Press, Ithaca.

Steg, L. and Vlek, C. 2009. Encouraging Pro-Environmental Behaviour: An Integrative Review and Research Agenda. *Journal of Environmental Psychology, Environmental Psychology on the Move* 29, 309–317. doi:10.1016/j.jenvp.2008.10.004

SWAN Research. 2011. *Stated NRW (Non-Revenue Water) Rates in Urban Networks*. Smart Water Networks Forum, Walton-On-Thames, Surrey.

Syme, G.J., Shao, Q., Po, M. and Campbell, E. 2004. Predicting and Understanding Home Garden Water Use. *Landscape and Urban Planning* 68, 121–128. doi:10.1016/j.landurbplan.2003.08.002

Thames Water. 2015. *Final Water Resources Management Plan 2015–2040*. Thames Water, Reading.

Turner, S.W.D. and Jeffrey, P.J. 2015. Industry Views on Water Resources Planning Methods – Prospects for Change in England and Wales. *Water and Environmental Journal* 29, 161–168. doi:10.1111/wej.12102

United Nations. 2015. *Sustainable Development Goals* [WWW Document]. Sustainable Development Knowledge Platform, Department of Economic and Social Affairs. https://sustainabledevelopment.un.org/?menu=1300.

van den Berg, C. 2015. Drivers of Non-Revenue Water: A Cross-National Analysis. *Utilities Policy* 36, 71–78. doi:10.1016/j.jup.2015.07.005

Whipple, C. 2016. *15 Litres of Water* [WWW Document]. www.15litresofwater.com/.

Willis, R.M., Stewart, R.A., Giurco, D.P., Talebpour, M.R. and Mousavinejad, A. 2013. End Use Water Consumption in Households: Impact of Socio-Demographic Factors and Efficient Devices. *Journal of Cleaner Production, Special Volume: Water, Women, Waste, Wisdom and Wealth* 60, 107–115. doi:10.1016/j.jclepro.2011.08.006

Zhang, H.H. and Brown, D.F. 2005. Understanding Urban Residential Water Use in Beijing and Tianjin, China. *Habitat International* 29, 469–491. doi:10.1016/j.habitatint.2004.04.002

6 Sanitation

Introduction

In 2015 more than 2.4 billion people did not have access to improved sanitation. Improved sanitation includes various forms of pit latrines and ecological sanitation as well as flush toilets. 70% people without access live in rural areas, but 18% of people living in cities, that is, more than 700 million urban residents, did not have access to sanitation (UNICEF and WHO, 2015). The Millennium Development Goal to halve the proportion of people without access to sanitation by 2015 was not met. Sustainable Development Goal 6 aims to 'by 2030, achieve access to adequate and equitable sanitation and hygiene for all and end open defecation, paying special attention to the needs of women and girls and those in vulnerable situations' (United Nations, 2015). Provision of sanitation to those who do not have access in urban areas is complex and highly debated. Many advocates maintain that water-based sewerage systems are the most effective technology for achieving good public health outcomes, while others promote the benefits of alternative, waterless sanitation technologies (Luthi et al., 2011; Satterthwaite et al., 2015). This is a major and urgent challenge for engineering, urban planning, public health and the global community.

The construction of water-based sanitation systems in Europe, North America and other parts of the Global North, along with the provision of clean drinking water, is widely considered the greatest public health intervention since the Industrial Revolution (Ferriman, 2007). However, it has come at the cost of pollution of natural water bodies, high water demand for flushing, high energy demand for pumping and treatment, and loss of nutrients and other resources. The sustainability of water-based sanitation is questionable in the context of continued urbanisation, energy constraints and forecasts for reduced water resource availability.

Providing a safe, clean, private place to defecate, urinate, vomit and manage menstruation is vital for human dignity and wellbeing. Access to toilets, including public toilets, is the subject of important social and political campaigning, particularly for people with disabilities, transgender people, women and the elderly (Gershenson and Penner, 2009; Herman, 2013). Poorly maintained public or shared toilets may be sites of gender-based violence and antisocial behaviour (Travers et al., 2011). In the Global South, provision of clean, safe toilet facilities at schools has been shown to improve attendance, particularly amongst girls (Freeman et al, 2012). The loss of public toilets and lack of accessibility to toilets in cities in the Global North restricts the ability of women, elderly people and those with medical conditions or disabilities to engage in public life (Greed, 1996). Fair, safe and decent provision of toilets and sanitation infrastructure is therefore a prerequisite to a modern, open and egalitarian society.

This chapter describes the evolution of water-based sanitation as the norm in the Global North and explores challenges to this model of infrastructure, including resource constraints and affordability. It then presents developments in waterless sanitation and analyses the challenge of sustainable sanitation from the perspective of the five frameworks of urban water sustainability.

Water-based sanitation

From a public health perspective, the purpose of sanitation infrastructure is to control microbes (Mara and Horan, 2003). In well-functioning sanitation infrastructure, disease-causing bacteria in waste are safely contained, separated from human contact and removed from the immediate urban environment. Helpful bacteria and protozoa then break down the waste for safe disposal or reuse. Municipal wastewater treatment is fundamentally a microbial process, as are composting, anaerobic digestion and most dry sanitation techniques. Just as some bacteria in the human body cause disease while others are essential for good health, microbes are both friend and foe for urban sanitation engineers.

Despite being at the heart of urban sanitation, bacteria were largely unknown in the mid-nineteenth century when the water-based sewerage model that has become the norm was established (Melosi, 2008). Early microscopes allowed people to see 'animicules' or tiny forms of life in water, but the bacteriological theory of disease was not widely accepted until the end of the century (Halliday, 1999). Disease outbreaks in rapidly urbanising cities were linked to poverty and poor living conditions (Hamlin, 1998). Proponents of contagionist theories claimed that disease was transferred between sick individuals and non-contagionists proposed that disease was caused by environmental conditions (Tarr, 1979). The non-contagionist miasma theory held that disease was caused by bad smells, including 'sewer gas', emanating from rotting matter. Water was thought to be 'self-purifying', so that disease-causing waste diluted in rivers or streams would eventually be made safe through natural processes. Water was a convenient medium to contain, dilute and transport rotting, smelly waste, alleviating the risk of disease (Tarr et al., 1984).

Prior to the nineteenth century household waste in cities in Europe and North America was contained in privy-vaults or cesspits (Gandy, 1999; Melosi, 2008; Tarr et al., 1984). These would be routinely emptied by 'nightsoil men' providing a private service or contracted by local authorities (Halliday, 1999). The waste collection typically occurred during the night to avoid carting dangerous, smelly 'nightsoil' through the streets during busy waking hours. Waste was routinely transferred to fields surrounding cities and used as fertiliser for agricultural and horticultural production. Surface water drains, where they existed, prevented localised flooding, directing relatively clean runoff to local waterways. In London, Boston, Paris and other places it was illegal for households to connect to the surface water sewers, an environmental regulation which prevented pollution of local streams and rivers and blockage of sewers with household waste (Halliday, 1999; Tarr et al., 1984).

During the nineteenth century, existing sanitation systems were under pressure from rapid population growth, dense urbanisation and industrialisation. Cesspits and privies filled more quickly as they served more people. Piped water supply also increased pressure on drainage and waste disposal systems. Where houses had previously obtained water from local wells or mobile water carts, piped water was available for the first time on a large scale (Halliday, 1999). Private companies and municipal authorities connected houses to new piped water supply, but without adequate provision for draining this additional water away from homes and streets (Tarr et al., 1984).

From the middle of the nineteenth century the water closet, or flushing toilet, became a popular device in homes concerned with good health, cleanliness and the latest technological advances. Water has been used for flushing waste since ancient times, but the modern flushing toilet has its origins in the eighteenth century. In 1778, Joseph Brammah registered a patent for a water closet in England, and by 1797 he had produced more than 6,000 of them (Halliday, 1999). George Jennings, Charles Crapper and other entrepreneurs in London and elsewhere improved on Brammah's design and promoted the water closet as a new household device, capitalising on Victorian concerns with cleanliness and the emergence of the bathroom as a new feature of private houses (Penner, 2014). Water closets were promoted as a means of washing away rotting matter that caused bad smells and disease, utilising the self-purifying property of water as a means of improving public health. Privies and cesspits were unable to cope with this additional volume of wastewater, and houses were for the first time allowed to connect to the public drains. Dumping waste into local waterways and streams, either directly or by flushing down drains that were originally designed for surface water, contributed to rapid decline in urban environments and public health. Overflowing cesspits, now filled with water, also contaminated shallow groundwater which supplied well water, further increasing risks to public health through the transmission of bacteria and viruses. As the volume of waste and wastewater increased, environmental conditions in cities declined, providing the basis for widespread disease, including epidemics of diseases such as cholera, typhoid and yellow fever (Halliday, 1999; Melosi, 2008; Tarr et al., 1984).

The second half of the nineteenth century saw the beginning of large-scale public works in response to environmental and public health crises. Removing waste changed from being a labour-intensive process employing night soil men and scavengers to a capital-intensive system involving the construction of vast networks of sewers (Allen, 2008; Tarr et al., 1984). In cities such as London and Paris the new wastewater infrastructure was retrofitted to surface water drainage networks that had evolved over previous decades and centuries (Gandy, 1999; Hamlin, 1998). In newer cities in North America which didn't already have surface water drains, combining surface water and wastewater in the same system provided economic and logistic efficiencies, particularly considering that surface water from streets was increasingly polluted with animal faeces, particularly from horse traffic (Tarr, 1979). Combining surface water and wastewater from homes and toilets was a cost-effective solution that minimised disruption to the city.

Water-based sanitation in general, and combined sewers in particular, were not adopted without criticism or debate about alternatives. The dilution of waste with water made it more difficult to transport and use as fertiliser (Halliday, 1999; Hamlin, 1992). Combining surface water and wastewater in the same pipe network required larger pipes than for separate systems and increased the risk of sewer gases escaping, supposedly causing the spread of disease. The earth closet was patented as an alternative to the water closet, including a device by Reverend Henry Moules from Shropshire in England, as well as numerous adaptations and improvements by US- and British-based inventors (Penner, 2014). In Amsterdam a pneumatic sanitation system, which transported waste using pressurised air, was installed in the 1870s (Geels, 2006; Luthi et al., 2011). In Manchester and other cities in England a pail system was implemented, as an improvement on the traditional night soil collection arrangements, with regular, systematised public collection of standardised waste containers (Penner, 2014). Colonel George E. Waring promoted separate, rather than combined, sewage systems throughout the US, notably implementing his design in Memphis after a series of outbreaks of cholera and yellow fever (Tarr, 1979). Despite concerns about wastage of resources and environmental impacts at the time, water-based sanitation proved an effective means of reducing public health risk and improving the quality of the immediate urban environment and became the norm for sanitation infrastructure in the Global North and colonial centres.

Water-based sanitation removed waste from inner cities but displaced environmental harm downstream, leading to continued efforts to develop technologies to treat wastewater before discharge. In the late nineteenth century this focussed on establishing sewer farms to utilise the nutrient-rich sewage for irrigation, with soils assisting purification of the waste (Halliday, 1999; Schneider, 2011). Knowledge about the role of bacteria and microbes in the purification of sewage informed innovations in the 1890s in England and the US, including the 'cultivation filter', the trickling filter and the septic tank. In 1906 German engineer Karl Imhoff patented the Imhoff tank, which combined anaerobic conditions for bacteria to treat wastewater with settling of solids (Schneider, 2011).

The most significant technological development in wastewater treatment was the invention of the activated sludge process at Manchester University in 1914 (Schneider, 2011). Building on work undertaken on aerobic bacteria and sewage treatment at the Lawrence Experimental Station in Massachusetts, Gilbert Fowler and Ernest Moore Mumford discovered that sewage sludge could be 'activated' by bubbling with air and reinoculation with older sludge that had previously been exposed to similar treatment. The microbial community within sewage was modified to dramatically increase the rate of degradation of organic solids and nitrogen compounds, providing the scientific basis for the development of modern sewage treatment processes.

Sewage treatment methods

The invention of the activated sludge process revolutionised sewage treatment and remains the basis for wastewater treatment in most major cities, supplemented with unit processes for screening, settling, clarifying and other physical and chemical techniques for removing pollutants from water (Figure 6.1) (Droste, 1997). Pumping air through wastewater and pumping water itself requires energy, and as environmental requirements for

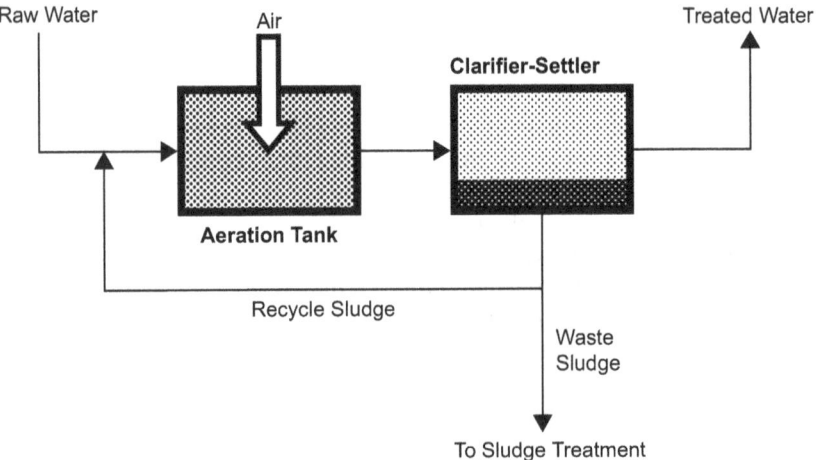

Figure 6.1 Activated sludge process

pollutant removal have tightened, the energy intensity of wastewater treatment has increased. Removal of nutrients such as nitrogen and phosphorus, and micro-pollutants such as pesticides and pharmaceuticals, requires more intensive physical and chemical treatment, further increasing the resource intensity of wastewater treatment processes (Joss et al., 2008). Improved efficiency of wastewater treatment focusses on optimising unit processes and pumping, and resource recovery, including using sludge as soil conditioner and recovering methane gas from anaerobic digestion of sludge for use as an energy source.

Less energy-intensive processes for wastewater treatment, including anaerobic digestion, septic tanks, stabilisation ponds and artificial wetlands, usually require more space and higher ambient temperatures than conventional processes, and are most applicable in warmer climates and for smaller towns and settlements (Hao et al., 2010; Luthi et al., 2011; Mara and Horan, 2003). Pollution risks are higher for distributed systems such as septic tanks, French drains and stabilisation ponds. These may only be suitable in rural or peri-urban areas with low population density as levels of treatment and control of discharges are less reliable than for centralised wastewater.

Challenges for water-based sanitation

After decades of modification, including dramatic improvement in wastewater treatment technologies, water-based sanitation systems have proved effective in eliminating disease, but concerns about resource loss, wastewater discharge and water scarcity remain. Toilet flushing is a significant contributor to urban water demand, combined sewer overflows are a major source of pollution, and resource recovery and reuse from sewage are increasingly important to urban sustainability. Extensive water-based sanitation systems are also expensive to build and complex to manage, particularly in rapidly urbanising cities in the Global South.

Toilet flushing accounts for around 20–30% of indoor household water consumption in most cities with water-based sanitation systems (see Chapter 5). The volume of toilet cisterns has reduced in recent decades as a water conservation measure. Dual-flush toilets are now typically 3 litres and 6 litres per flush, with some models down to 2 and 4 litres, compared to older single flush toilets with flush volumes of 13 litres or more (Grant and Moodie, 2002; Keating and Styles, 2004; Schlunke et al., 2008). Ultra-low-flush toilets are emerging onto the market, such as the Propelair toilet invented in the UK, which has a flush volume of 1.5 litres and is assisted by pneumatic pressure (Propelair, 2017).

Combined sewers built to transfer both wastewater and surface water are now a major source of environmental pollution in cities in the US and Europe (Novotny et al., 2010). Combined sewers are designed to overflow directly into rivers and streams during heavy rainfall events. During heavy rainstorms, large volumes of surface water flow into the sewers, until they reach capacity (Butler and Davies, 2010). The purpose of 'combined sewer overflows' (CSOs) is to prevent surface water flooding or uncontrolled sewer flooding in streets and homes. CSOs are designed so that relatively dilute sewage overflows into rivers rather than flooding homes and streets. As cities have become more urbanised and previously permeable surfaces have been paved over, more rainfall runs off into the sewers and CSOs occur with a greater frequency than when the systems were first designed. Discharge of untreated sewage from CSOs into urban rivers is a major environmental problem, which is discussed further in Chapter 7.

In rapidly urbanising cities of the Global South, implementation of extensive water-based sewerage networks is further hampered by lack of access to capital to pay for construction and governance and administrative capacity to operate and maintain such large-scale systems (Luthi et al., 2011). Households in suburban and peri-urban areas of cities in the Global South may install decentralised sanitation systems, using private groundwater wells to provide water for toilet flushing, and septic tanks and soakaways to dispose of wastewater. This may provide appropriate access to sanitation, but it increases risks to the local environment and public health through over-abstraction of water and microbial contamination of groundwater, particularly as population density increases.

Small bore sewers, localised networks and decentralised sewage treatment plants provide further lower-cost options for water-based sanitation infrastructure, particularly for retrofitting to dense, informal urban settlements. Sharing the cost, administration, design and construction of sewerage systems between local residents and local government can provide greater success at lower overall cost, as has been the case for the Orangi Pilot Project, which started in Karachi in the 1980s and has expanded to other cities in Pakistan (Hasan, 2006, 2002).

Providing flushing toilets and sewer connections without sewage treatment displaces environmental and health risk downstream, in a similar way to early nineteenth-century sewage systems in Europe and the US. Baum et al. (2013) estimated that in 2010 only 40% of the global population had access to improved sanitation with treatment, compared to official UN statistics of 62% with access to improved sanitation (with or without treatment). Appropriate sewage treatment requires additional investment beyond provision of toilets and sewers, and it requires reliable electricity supply for pumping and treatment.

Sludge disposal is a common challenge for sanitation infrastructure. Disposal of sludge from wastewater treatment works in the Global North has changed over recent decades from dumping at sea to disposal on land and burning for energy. Sewage sludge has been marketed as a fertiliser and soil conditioner and applied in bulk on farms as well as sold to household gardeners (Schneider, 2011). Using sludge as fertiliser may be a sustainable source of nutrients for agriculture, but it has also raised concerns about pollution. Wastewater treatment works often treat industrial as well as household waste, and heavy metals, pesticides, pharmaceuticals and other contaminants may be present in urban wastewater and surface water. These contaminants can become concentrated in the sewage sludge as they are removed from the water, and are a source of safety concern for land application, particularly for farm workers and local residents, but also for food production. Land application of sewage sludge is banned in Belgium and other European countries but is widely practiced in the US and UK. The risk of pollution depends upon the nature of the wastewater, which is determined by the activities and materials used in the city, as well as treatment and application of the sludge or resulting fertiliser product. Controversy over sludge reuse highlights the complexity of trade-offs in urban sustainability, between managing risks of pollution and the benefits of resource recovery. It also demonstrates the interconnection between environmental systems in cities. Discharging industrial waste to sewers may reduce pollution risk in the immediate urban environment, only to redistribute pollutants and health risk to sewage treatment and sludge reuse systems.

Waterless sanitation

The goal of universal water-based sanitation has been widely questioned and is considered by many to be unachievable, unsustainable and undesirable (George, 2009; Langergraber and Muellegger, 2005). Water-based sewerage systems require large-scale capital investment, high standards of governance, robust financing for operating and maintenance, and continuous supply of water and electricity. In many cities, the elements required for successful, centralised water-based infrastructure do not yet exist. Growing awareness of limits to water resources and the environmental impacts of wastewater treatment and effluent discharge have also led to propositions for waterless sanitation in both the Global North and South.

Echoing nineteenth-century alternatives such as the pail system and the earth closet, recent movements in ecological sanitation emphasise resource recovery, decentralisation and the elimination of the use of water to transfer dangerous waste (Penner, 2014). Dry sanitation can be as simple as a pit latrine, but also includes sophisticated composting toilets and innovative resource recovery systems. Dry sanitation systems avoid demand for water and the need for energy-intensive wastewater treatment. Waterless and ecological sanitation systems also improve options for recovery of energy and nutrients from the waste, compared with recovery from wastewater (Luthi et al., 2011). However, they require safe systems for collecting, emptying and treating waste, and significant challenges remain in developing reliable, affordable systems on a scale that is required to meet the global challenge of urban sanitation.

Ecological sanitation has mostly been implemented in rural areas. Outhouse superstructures can be built over latrine pits and moved when the pit is full, leaving the waste to decompose so that it can be used as a fertiliser in nearby fields or gardens. Various designs for latrines and ecological sanitation systems have been developed, with an emphasis on locally available materials, use of local labour and engagement with local communities.

The Community-Led Total Sanitation (CLTS) movement began in Bangladesh and provides a structured methodology for engaging communities in the design and construction of their own ecological sanitation systems (Chambers, 2009; Kar and Chambers, 2008). CLTS begins by demonstrating the dangers of open defecation and requires community commitment to eliminate it, before designing and constructing latrines using community resources. It has been criticised for its emphasis on shaming individuals who openly defecate and placing responsibility on communities rather than governments to provide basic services (Bartram et al., 2012). CLTS has been most widely applied in rural settings, and it has faced challenges in transferring the method to urban contexts with higher density and more dynamic community structures.

The application of waterless sanitation to urban areas is more difficult than in rural areas. Space constraints makes pit emptying more difficult and dangerous, and higher population densities place additional pressure on the local environment and increase public health risks. Addressing the challenge of devising a viable ecological sanitation system for urban areas has become a significant focus of design and development efforts. Pit latrines in urban settings are usually associated with slums and informal settlements but may also be found in suburban and peri-urban areas without access to sewerage networks or space and resources for septic tanks. Well-designed pit latrines are lined to prevent contamination of groundwater, have a secure slab above the pit, are well ventilated to reduce smells and have superstructure that allows safety and privacy of use. Pit latrines that are poorly designed and built and not properly maintained can become a source of health risk and environmental pollution, rather than a solution.

Disposal of the contents of pits is a significant challenge in cities. Pit sludge may be officially disposed of at municipal wastewater treatment facilities, or buried or dumped informally and illegally, displacing the health and environmental hazard. Emptying pits can be dangerous work, made more difficult by the disposal of general waste, including household plastics, condoms and menstrual products, in addition to organic excreta. The labour intensity of latrine emptying and the dangers of disposal of sludge are reminiscent of the challenges of the privy-vault and cesspool systems in nineteenth-century cities of the Global North. Attention to the challenges of building, maintaining and emptying latrines and managing sludge has focussed on both technical and behavioural change and innovative business models. Development agencies, NGOs and local entrepreneurs have devised pumps and carts to make pit emptying easier and safer, as well as latrines, including the 'Gulper' manual pump system and the UN-HABITAT Vacutug vacuum pump (Thye et al., 2011). Disposal of sludge from pit latrines in the Global South is a more direct health and pollution risk than disposal of sewage sludge in the Global North, as untreated waste material may be dumped directly into the environment. Depending on local conditions, this can either displace the health risk or magnify it, as waste that was previously contained in a latrine is exposed openly to the environment. Methods for treating sludge focus on stabilisation and resource recovery, including anaerobic digestion, treatment through conventional wastewater treatment works, and composting.

Urine diverting toilets (UDTs) enable source separation of urine and faeces to improve sludge management and resource recovery. The municipality of eThekwini in Durban, South Africa, installed 80,000 UDTs in peri-urban households between 1999 and 2010 (Grounden et al., 2006; Luthi et al., 2011). The toilets were designed with two vaults to contain the solid waste and a pedestal that can be moved between the two (Figure 6.2). While one vault is in use, the waste in the other is drying and stabilising, to facilitate safe and easy pit emptying. The pits are designed to alternate every one to two years. Urine is diverted from the pits by the installation of a urinal and a separator within the bowl of the pedestal. The urine soaks into the ground, to reduce moisture and odour in the solid waste pits. Urine separation has also been implemented in Europe to reduce toilet flushing volume and to improve opportunities for resource recovery (Blume and Winker, 2011). Urine has a high concentration of nutrients compared to faeces, and future developments of UDTs include recovery of nutrients from urine for fertiliser production.

Figure 6.2 Urine diverting toilets, eThekwini Municipality, South Africa

As interest in new toilet technologies and systems builds in the Global South, the possibility of long-term reform of urban sanitation in the Global North is also being considered. The market for high-tech composting toilets has grown rapidly in the Global North, particularly for holiday homes and 'off-grid' settlements. Such systems can involve electric heating to promote bacterial composting activity and fans for ventilation, reducing smell, drying the waste and promoting rapid composting. Waterless urinals are also installed in men's bathrooms in the Global North to eliminate water for flushing, but these require a change in cleaning and maintenance practices, and can involve disposable cartridges, which leads to a shift in environmental impacts (Hills et al., 2002).

Intermediate container-based systems that enable safe collection of waste and treatment to recover energy and nutrients in localised facilities have been developed for implementation in dense urban areas. In the Global North these systems serve communities that don't have access to water-based sewerage systems, such as allotment gardeners, events, including music festivals, that typically use portable toilets, and those living on boats in canals and marinas.

In the Global South container-based initiatives for urban ecological sanitation include the Kenyan social enterprise Sanergy, which manufactures stand-alone toilets with removable cartridges that are collected and emptied into biodigesters to produce renewable energy and fertiliser. Sanergy aims to develop a viable business model through franchising their user-pays 'Fresh Life Toilets' and provision of sanitation technology and services, together with nutrient and energy recovery. By 2017 Sanergy had installed more than 1,000 toilets (Sanergy, 2017).

London-based designer Virginia Gardiner developed the Loowatt system, with a waterless toilet lined with biodegradable plastic, and removable waste cartridges delivered to local biodigester facilities to produce methane and further composting to produce fertiliser (Figure 6.3) (Loowatt, 2017). Loowatt's system serves 100 households in Antananarivo, Madagascar, and serves various festivals across the UK.

Waterless sanitation seems unlikely to replace water-based systems in the Global North in the immediate future, and the emphasis on dry sanitation as the preferred technology for the Global South remains controversial. Dry sanitation is seen by some as an inferior and risky alternative to water-based systems, which have been shown to deliver good public health outcomes in the Global North. Low coverage of water-based toilets and sewerage systems is a political and economic rather than technical failure. For some development professionals, the focus on waterless sanitation as a technical solution diverts attention away from failures of urban governance by proposing a technically inferior solution for the urban poor compared to the water-based sanitation used by wealthier households and cities in the Global North. Alternative business and technology models such as Sanergy and Loowatt hold promise, but have remained small-scale compared to the global challenge.

Figure 6.3 Loowatt waterless, container-based toilet in Madagascar

(Source: Loowatt, used with permission)

Framing sanitation

Sanitation is the most basic and essential of urban infrastructure services, and yet it is fraught with complexity, controversy and persistent failure. Sanitation systems manage the most abject wastes arising from deeply private bodily functions. Discussions about sanitation are too often avoided because of disgust and cultural taboo. Framing sanitation as an issue of basic human rights and dignity, an environmental problem, a source of resources and the foundation for good public health allows professional and public discussion to move beyond queasiness. The form of sanitation, inequalities in provision, environmental impacts, and challenges of finance and management demonstrate its political, social and cultural dimensions, alongside technical and scientific developments. The frameworks of sustainable development, ecological modernisation, socio-technical systems, urban political ecology and radical ecology each highlight different aspects of the sanitation challenge, from both complementary and contradictory viewpoints.

Sustainable development

Sanitation has lagged behind water provision in global debates and initiatives for sustainable development. It was a late inclusion to the Millennium Development Goals and progress was weaker than for water, with 32% of the global population lacking access in 2015, compared with the target of 23% (UNICEF and WHO, 2015). The Sustainable Development Goals make much stronger recognition of the importance of sanitation and have a more ambitious target of achieving universal access to improved sanitation by 2030. Provision of sanitation is better in urban than rural areas, but rapid urbanisation provides additional challenges and provision risks falling behind population growth in many cities (Satterthwaite et al., 2015).

Development policy and practice recognises the need for a variety of technical options for sanitation, to meet the Sustainable Development Goal of safe, dignified, universal provision. Sanitation provision has been conceived as a ladder, with open defecation at the bottom and improved sanitation for individual households at the top (Table 6.1).

Provision of improved sanitation remains the primary concern for sustainable development, but environmental protection, water efficiency and resource recovery are also important factors in achieving sustainability. Ecological sanitation and more recent developments such as Loowatt and Sanergy are attractive as a means of conserving water resources, avoiding large-scale infrastructure construction and recovering nutrients and energy. However, their application in urban areas is constrained by the challenges of finding space for latrines in dense urban settlements and safe management and disposal of sludge.

Water-based sanitation systems remain important options for sustainable development but without appropriate sewage treatment they may contribute to environmental degradation and pose serious health risks. Sustainable Development Goal 6 recognises the need for treatment of sewage and sludge to protect environment and health, requiring stronger attention to infrastructure systems as well as individual toilets or latrines. Creating institutions capable of delivering and managing sanitation systems, whether

Table 6.1 The sanitation ladder (WHO and UNICEF, 2017)

Safely managed	Use of improved facilities which are not shared with other households and where excreta are safely disposed in situ or transported and treated off-site. Improved facilities include: – Flush or pour flush toilet discharging to public sewer or septic tank – Pit latrine with slab – Ventilated Improved Pit latrine (VIP) – Composting toilet
Basic	Use of improved facilities which are not shared with other households.
Limited	Otherwise acceptable sanitation facilities that are shared between two or more households.
Unimproved	No hygienic separation of excreta from human contact. Includes latrines without slab or platform, hanging latrines and bucket latrines.
Open defecation	Human faeces disposed of in fields, forests, bush, water bodies, beaches or other open areas, or with solid waste

water-based or dry, centralised or decentralised, is a considerable challenge for sustainable development.

In the Global North nutrient recovery and environmental protection remain challenges for sustainable cities. Energy-intensive treatment processes, sludge contaminated with industrial waste, and the complexity of pollutants in wastewater are constraints on the capacity of the sector to deliver sustainability even in contexts where basic public health goals have been met. Combined sewer overflows are a source of serious environmental pollution leading to a range of responses from capital-intensive interceptor tunnels to distributed green infrastructure, which are discussed in detail in Chapter 7.

Ecological modernisation

In 2011 the Bill and Melinda Gates Foundation announced a high-profile competition for universities and entrepreneurs to 're-invent the toilet' (Gates Foundation, 2012). The competition brief was to design a stand-alone system that is not dependent on connections to electricity, water or a septic system, does not discharge pollutants and costs no more than 5 cents per day to run. Just as Bill Gates set out to put personal computing within reach of ordinary people through innovative software engineering and business development, so the reinvent the toilet competition proposes to transform not only the technology of sanitation but also the way it is funded and operated. 'Reinvent the toilet' typifies an ecomodernist approach to sanitation. It seeks to promote technological innovation as the means to solving a social and environmental crisis. It recognises the importance of economics and affordability in delivering viable outcomes, and it is focussed on individual households and private business models rather than shared infrastructures.

The Gates Foundation has provided funding to support Loowatt and Sanergy, both of which promote self-sufficient business models, based on a fee for service, operated by small-scale entrepreneurs rather than large municipal institutions. This sanitation entrepreneurial approach represents a fundamentally different set of relationships between citizens and infrastructure. The decentralised infrastructure corresponds to decentralised, privatised ownership and operation. People living in cities are conceived as customers of a sanitation service, rather than citizens with a right to good public health and clean environment.

Ecological sanitation, particularly as presented by the entrepreneurial model, provides opportunities for social enterprise and local economic activity based on provision of sanitation and resource recovery services (Schaub-Jones, 2011). Regulation of such enterprises requires different governance arrangements to conventional systems, with the possibility for more actors to be involved. In the Global North this will also require changes to public health regulations and planning, to allow for urban development served by ecological sanitation rather than requiring connection to sewer.

Implementing ecological sanitation systems with service providers empty-ing toilets may also lead to reconfiguration of domestic and urban space. Just as the removal of privies and the introduction of indoor flushing toilets changed the layout of urban and suburban housing, so too might houses and streets be reorganised once more to accommodate the shift towards ecological sanitation. Integrating sanitation into dense informal settlements and slums has also changed the layout of streets and led to wider goals of upgrading infrastructure and the urban environment (Parikh et al., 2013).

Ecological modernisation can also be seen in the attention to resource recovery and efficiency in wastewater treatment in the Global North. Changes in unit processes and treatment systems to optimise energy con-sumption, recover energy through methane production and burning sludge, and recover nutrients in sludge are examples of how technological develop-ment and innovation can reduce impacts on resources and the environment. Regulation has an important role in driving water treatment discharge standards, which provides incentives to improve the efficiency and effective-ness of treatment. Regulation may also be a barrier to resource recovery, as concerns about public health and pollution limit application of sludge to land, and ever-tighter discharge standards drive up energy requirements for wastewater treatment.

Socio-technical transitions

In the history of cities, water-based sanitation is a relatively recent infra-structural norm. For centuries, cities in Europe, Asia and other parts of the world used waterless systems of night soil collection and nutrient recycling. And yet, it now seems unimaginable for a modern city to have any other form of sanitation. The flushing toilet and city-scale sewerage system have become almost universal signifiers of modern urban life. Water-based sani-tation systems are effectively 'locked-in' to cities in the Global North, and into cultural and institutional definitions of what a modern, safe sanitation system should be.

Infrastructure lock-in involves institutional and cultural norms as well as the technologies and networks that are embedded in cities. In London, the completion of the intercepting sewers project in the mid-nineteenth cen-tury required the formation of a centralised authority to finance, design and construct sewerage infrastructure across the whole city (Halliday, 1999). Through various changes in ownership and governance of water and san-itation infrastructure the institutional scale of management of sanitation infrastructure has remained at the city or region. This institutional lock-in serves to reinforce technological lock-in in cities. Decentralised sanitation technologies are not obviously consistent with the dominant institutional structures for sanitation infrastructure, which have developed to support centralised, water-based sanitation systems.

The 'flush and forget' experience of water-based sanitation has become a cultural norm in modern cities. Nineteenth-century aversion to bad smells and miasma has become embedded in the flushing toilet and associated sewerage systems. Fear of contamination, disgust at bad smells and the challenges of servicing, including emptying and disposal, of the waste are cultural as well as institutional barriers to alternative sanitation technologies.

Inappropriate use of toilets or latrines as general waste disposal units is a global challenge for sanitation engineers and workers. Avoiding latrine pits or composting toilets filling with non-organic waste such as plastic bags, condoms, sanitary napkins, nappies and other rubbish requires ongoing education and engagement of users. Flushing these and other items, such as wet wipes, down toilets in the Global North creates blockages in sewers, pumps and wastewater treatment works. Pouring fats, oils and greases down sinks and drains is a major source of sewer blockage and the focus of behaviour change campaigns. Sewer blockages can cause local sewer flooding, with obvious risks to public health. Pit latrines with non-degradable waste also increase public health risks, as bacterial decomposition is interrupted and pits fill faster and are more difficult to empty. The social and cultural association of toilets with easy disposal of abject waste is a somewhat inevitable outcome of the role that the toilet has evolved to play in cities, yet it is a source of considerable problems for system managers and engineers.

The history of sanitation shows that the water-based infrastructure model has always been contested (Hamlin, 1992; Tarr et al., 1984). The persistence of alternative systems and technologies indicates a niche role for waterless sanitation, but further expansion requires higher-level changes in politics, investment and culture. The success of alternative and decentralised technologies depends upon institutional and economic reform, as much as the successful invention of a desirable, affordable and functional waterless toilet.

A socio-technical analysis of sanitation systems can also help to reveal the interdependence of different infrastructure systems and the implications of changing social and political expectations. Social and political demand for higher-quality wastewater discharge to the environment has driven existing wastewater treatment works to more energy-intensive treatment processes, driving up demand for energy, which has further detrimental environmental impacts.

Political ecology

The failure to deliver safe, dignified sanitation to more than 700 million people living in cities is shameful. It is one of the starkest indicators of the inequalities of global development. Without sanitation infrastructure, the poorest people in cities in the Global South live in the most degraded and dangerous environmental conditions.

In most cities of the Global North, sanitation infrastructure was constructed by the state and continues to be owned and managed by public authorities. Sewerage infrastructure has been privatised in the UK, and the private sector plays a significant role in sanitation services around the world. However, for most cities, sanitation is a critical infrastructure service provided by the state to ensure universal access and good public and environmental health. Sanitation infrastructure, even where it is provided by private firms, requires strong regulation and governance to ensure fairness and safety. Development policies for the Global South that emphasise privatisation and deregulation of infrastructure have struggled to address the scale of the challenge of sanitation in rapidly urbanising cities.

In Cape Town, South Africa, uneven provision of sanitation has become a highly contentious political issue (McFarlane and Silver, 2017). In 2011 residents from informal settlements protested at the provision of shared toilets, some without walls, and others using container-based systems that were not emptied or maintained regularly. Protests included dumping toilets and their contents onto a major motorway, but also included organised advocacy aligned with other housing and human rights issues. Sanitation in South Africa reflects the complex politics of the post-apartheid era, but it also points to wider patterns of inequality in infrastructure provision in cities around the world. The Cape Town protesters were campaigning for access to flushing toilets, not the inferior decentralised, dry sanitation systems that had been provided. The 'bucket' toilets were reminiscent of apartheid-era sanitation provision to black settlements, and represented not only infrastructural failure but the failure to address inequality and racial injustice in the new South Africa. Despite the potential benefits of well-designed and managed waterless, container-based sanitation highlighted by Sanergy and Loowatt, the experience of poor, black Cape Town residents was of dangerous, smelly, disgusting toilets, in stark contrast to the flushing toilets and sewers in wealthier neighbourhoods.

Inequality in provision of sanitation in cities in the Global South is in part a result of rapid urbanisation, but also reflects longer histories of colonialism. McFarlane's analysis of sanitation in Bombay/Mumbai compares the contemporary efforts to clean up the urban environment through demolition of slums and the creation of enclaves for wealthy elites to similar efforts by the colonial administration in the nineteenth century (McFarlane, 2008). Recent programmes to provide sanitation to slums have focussed on shared toilet blocks, compared to nineteenth-century efforts to replicate the model of city-wide sewerage provision from English cities. Governance and provision of sanitation was as fragmented and reflective of dominant power structure in nineteenth-century Bombay as it is in contemporary Mumbai.

The political ecology of sanitation helps to reveal the power relationships at play in the provision or failure of infrastructure. Whilst nineteenth- and twentieth-century sanitation engineering and administration emphasised the collective importance of good public health through universal provision, more recent trends towards fragmentation of infrastructure provision, underpinned by privatisation and deregulation, protect wealthier enclaves from the health and environmental risks experienced by the urban poor. Technical alternatives to water-based sanitation may provide options to improve urban environments and health, but unless democratic and economic inequalities are addressed it is likely that they will continue to be experienced as inferior forms of infrastructure that exemplify wider urban divisions.

Radical ecology

The politics of sanitation has also been a strong theme within the environmental movement, particularly in relation to the appropriate and alternative technology movements. Reducing water wastage in toilet flushing and recovering nutrients from waste were strong themes in engineering and design responses to the environmental crisis in the 1970s. These technical programmes were also political, aiming to develop 'autonomous houses' that not only reduced water wastage, recycled nutrients and eliminated pollution, but also transformed society (Penner, 2013). For radical ecologists, composting toilets, biodigesters and other off-grid sanitation technologies are practically and symbolically at the heart of the transformation to a sustainable society.

Dry sanitation assumes water is a scarce resource that should not be contaminated with human waste. The recovery of energy and nutrients reflects a broader concern for resource efficiency and reducing pollution of the environment. This has driven more recent attention to the role of ecological and sustainable sanitation in urban metabolism and the circular economy (Luthi et al., 2011). Radical designers and engineers in the 1970s focussed on autonomous houses, disconnected from urban infrastructure. Contemporary proponents of alternative sanitation systems focus more on communal or networked options for managing waste and recovering nutrients as part of wider transformations to sustainable cities.

Ecological sanitation also represents a shift in the relationship between individual bodies, technology and the environment. In modern cities individuals flush away their waste, for large utilities to take care of. Modern citizens unthinkingly delegate responsibility for the impact of their basic bodily functions on the environment to large, industrial infrastructure systems. These impacts occur at a distance and are mediated by large and complex sociotechnical systems. In ecological sanitation systems the processes of emptying pits or containers and composting or biodigesting the waste are more immediate and local. This has the potential to reconfigure the relationship between human bodily functions and environmental systems, reconnecting citizens and communities to ecological cycles of nutrients, water and waste.

Sanitation in sustainable cities

Access to sanitation is a basic human right (United Nations General Assembly, 2010). A fair, clean, safe system for delivering sanitation without polluting or over-exploiting the environment is a basic prerequisite for a sustainable city. Sanitation is too often overlooked in debates about urban sustainability, either because it is taken for granted in cities in the Global North or because it is an intractable, dirty problem without simple solutions in the Global South. Sanitation is the infrastructure that deals with private and disgusting bodily functions, and without it modern urban life is intolerable. In sanitation infrastructure bodily metabolisms directly connect to urban metabolisms, through flows of nutrients, water and energy.

The current domination of water-based forms of urban infrastructure is an outcome of particular historical circumstances in the rapidly urbanising cities of Europe and North America in the nineteenth century. Despite having its origins in scientific misconceptions, water-based sanitation works. Cities with well-functioning water-based sanitation infrastructure are largely free of waterborne disease and the smell of sewage. Flushing toilets and sewers enable standards of household and personal cleanliness and environmental protection that support good public health. Modern wastewater treatment systems clean the water used to transport waste and return it to the environment.

Water-based sanitation is effective, but it is not necessarily sustainable. It consumes around one-third of water used in homes, and it requires energy for pumping and treating wastewater. Recovering nutrients and energy from water is possible, though more complex than directly recycling from the waste itself. The capital intensity and technical complexity of water-based sanitation infrastructure have contributed to unequal construction and performance in the Global South.

Technical alternatives to water-based sanitation stem from traditional waste management, environmental design and modern entrepreneurialism. Controversies over poor provision of sanitation, such as the protest about container-based systems in Cape Town, demonstrate that the form of sanitation matters to people. Designing a clean, safe, odour-free dry sanitation system is a technical challenge, but it also requires cultural and institutional change for alternative sanitation systems to be accepted as equal to or better than flushing toilets.

Sustainable sanitation cannot be defined by a particular technology or system. Adapting existing infrastructure to reduce environmental impacts and improve resource efficiency and recovery will be important in achieving urban sustainability. New systems and technologies provide opportunities to more radically transform the relationships between people, bodies, infrastructure and environments, but the challenge remains to deliver safe, reliable, dignified sanitation to everyone living in cities, particularly the poor and marginalised.

References

Allen, M.E. 2008. *Cleansing the City: Sanitary Geographies in Victorian London.* Ohio University Press.

Bartram, J., Charles, K., Evans, B., O'Hanlon, L. and Pedley, S. 2012. Commentary on Community-Led Total Sanitation and Human Rights: Should the Right to Community-Wide Health Be Won at the Cost of Individual Rights? *Journal of Water and Health* 10, 499–503. doi:10.2166/wh.2012.205

Baum, R., Luh, J. and Bartram, J. 2013. Sanitation: A Global Estimate of Sewerage Connections Without Treatment and the Resulting Impact on MDG Progress. *Environmental Science & Technology* 47, 1994–2000. doi:10.1021/es304284f

Blume, S. and Winker, M. 2011. Three Years of Operation of the Urine Diversion System at GTZ Headquarters in Germany: User Opinions and Maintenance Challenges. *Water Science and Technology* 64, 579–586. doi:10.2166/wst.2011.530

Butler, D. and Davies, J. 2010. *Urban Drainage*, Third Edition. CRC Press, London.

Chambers, R. 2009. Going to Scale with Community-Led Total Sanitation: Reflections on Experience, Issues and Ways Forward. *IDS Practice Papers* 2009, 1–50. doi:10.1111/j.2040–0225.2009.00001_2.x

Droste, R.L. 1997. *Theory and Practice of Water and Wastewater Treatment.* Wiley-Blackwell, Chichester.

Ferriman, A. 2007. BMJ Readers Choose the "Sanitary Revolution" as Greatest Medical Advance Since 1840. *BMJ: British Medical Journal* 334, 111.

Freeman, M., Greene, L., Dreibelbis, R., Saboori, S., Muga, R., Brumback, B. and Rheingans, R. 2012. Assessing the Impact of a School-Based Water Treatment, Hygiene and Sanitation Programme on Pupil Absence in Nyanza Province, Kenya: a Cluster-Randomized Trial. *Tropical Medicine and International Health* 17(3), 380–391.

Gandy, M. 1999. The Paris Sewers and the Rationalization of Urban Space. *Transactions of the Institute of British Geographers* 24, 23–44.

Gates Foundation. 2012. *Reinvent the Toilet* [WWW Document]. Bill & Melinda Gates Foundation. www.gatesfoundation.org/What-We-Do/Global-Development/Reinvent-the-Toilet-Challenge.

Geels, F.W. 2006. The Hygienic Transition from Cesspools to Sewer Systems (1840–1930): The Dynamics of Regime Transformation. *Research Policy* 35, 1069–1082. doi:10.1016/j.respol.2006.06.001

George, R. 2009. *The Big Necessity: Adventures in the World of Human Waste.* Portobello Books Ltd, London.

Gershenson, O. and Penner, B. (Eds.) 2009. *Ladies and Gents: Public Toilets and Gender.* Temple University Press, Philadelphia, PA.

Grant, N. and Moodie, M. 2002. Focus – Water Closets: Best Practice Since the Water Fittings Regulations 1999. *GreenPro News* 12–14.

Greed, C.H. 1996. Planning for Women and Other Disenabled Groups, with Reference to the Provision of Public Toilets in Britain. *Environmental Plan A* 28, 573–588. doi:10.1068/a280573

Grounden, T., Pfaff, B., Mcleod, N. and Buckley, C. 2006. *Provision of Free Sustainable Basic Sanitation: The Durban Experience.* Presented at the 32nd WEDC International Conference, Colombo, Sri Lanka.

Halliday, S. 1999. *The Great Stink of London.* Sutton Publishing, Stroud, Gloucestershire.

Hamlin, C. 1992. Edwin Chadwick and the Engineers, 1842–1854: Systems and Antisystems in the Pipe-and-Brick Sewers War. *Technology and Culture* 33, 680–709. doi:10.2307/3106586

Hamlin, C. 1998. *Public Health and Social Justice in the Age of Chadwick: Britain, 1800–1854*. Cambridge University Press.

Hao, X., Novotny, V. and Nelson, V. 2010. *Water Infrastructure for Sustainable Communities*. IWA Publishing, London.

Hasan, A. 2002. A Model for Government-Community Partnership in Building Sewage Systems for Urban Areas: The Experiences of the Orangi Pilot Project – Research and Training Institute (OPP-RTI), Karachi. *Water Science and Technology* 45, 199–216.

Hasan, A. 2006. Orangi Pilot Project: The Expansion of Work Beyond Orangi and the Mapping of Informal Settlements and Infrastructure. *Environment & Urbanization* 18, 451–480. doi:10.1177/0956247806069626

Herman, J. 2013. Gendered Restrooms and Minority Stress: The Public Regulation of Gender and its Impact on Transgender People's Lives. *Journal of Public Management and Social Policy* 19, 65–80.

Hills, S., Birks, R. and McKenzie, B. 2002. The Millennium Dome "Watercycle" Experiment: To Evaluate Water Efficiency and Customer Perception at a Recycling Scheme for 6 Million Visitors. *Water Science and Technology* 46, 233–240.

Joss, A., Siegrist, H. and Ternes, T.A. 2008. Are We About to Upgrade Wastewater Treatment for Removing Organic Micropollutants? *Water Science and Technology* 57, 251–255. doi:10.2166/wst.2008.825

Kar, K. and Chambers, R. 2008. *Handbook on Community-Led Total Sanitation*. Institute of Development Studies and Plan International (UK), Brighton and London.

Keating, T. and Styles, M. 2004. *Performance Assessment of Low Volume Flush Toilets*. Southern Water, Worthing.

Langergraber, G. and Muellegger, E. 2005. Ecological Sanitation – A Way to Solve Global Sanitation Problems? *Environment International* 31, 433–444. doi:10.1016/j.envint.2004.08.006

Loowatt. 2017. *Loowatt Luxury Waterless Toilet Technology* [WWW Document]. Loowatt Luxury Waterless Toilet Technology. https://loowatt.com/.

Luthi, C., Panesar, A., Schutze, T., Norstrom, A., McConville, J., Parkinson, J., Saywell, D. and Ingle, R. 2011. *Sustainable Sanitation in Cities: A Framework for Action*. Papiroz Publishing House, Rijswijk, the Netherlands.

Mara, D. and Horan, N. 2003. *Handbook of Water and Wastewater Microbiology*. Academic Press, London and San Diego.

McFarlane, C. 2008. Governing the Contaminated City: Infrastructure and Sanitation in Colonial and Post-Colonial Bombay. *International Journal of Urban and Regional Research* 32, 415–435. doi:10.1111/j.1468-2427.2008.00793.x

McFarlane, C. and Silver, J. 2017. The Political City: "Seeing Sanitation" and Making the Urban Political in Cape Town. *Antipode* 49, 125–148. doi:10.1111/anti.12264

Melosi, M.V. 2008. *The Sanitary City: Environmental Services in Urban America from Colonial Times to the Present*. University of Pittsburgh Press.

Novotny, V., Ahern, J. and Brown, P. 2010. *Water Centric Sustainable Communities*. John Wiley and Sons, Hoboken.

Parikh, P., Parikh, H. and McRobie, A. 2013. The Role of Infrastructure in Improving Human Settlements. *Proceedings of the Institution of Civil Engineers – Urban Design and Planning* 166, 101–118.

Penner, B. 2013. *Bathroom*. Reaktion Books, London.

Penner, B. 2014. *Bathroom*. Reaktion Books, London.

Propelair. 2017. *Propelair® | The Toilet Reinvented* [WWW Document]. Propelair®. www.propelair.com/homepage/.

Sanergy. 2017. *Sanergy* [WWW Document]. http://saner.gy/.

Satterthwaite, D., Mitlin, D. and Bartlett, S. 2015. Key Sanitation Issues: Commitments, Coverage, Choice, Context, Co-Production, Costs, Capital, City-Wide Coverage. *Environment and Urbanization Brief* 31. International Institute for Environment and Development, London.

Schaub-Jones, D. 2011. Market-Based Approaches in Water and Sanitation: The Role of Entrepreneurship. *Waterlines* 30, 5–20. doi:10.3362/1756–3488.2011.002

Schlunke, A., Lewis, J. and Fane, S. 2008. *Analysis of Australian Opportunities for More Water Efficient Toilets*. Institute for Sustainable Futures, University of Technology Sydney, Sydney.

Schneider, D. 2011. *Hybrid Nature: Sewage Treatment and the Contradictions of the Industrial Ecosystem*. MIT Press, Cambridge, MA.

Tarr, J.A. 1979. The Separate Vs. Combined Sewer Problem. A Case Study in Urban Technology Design Choice. *Journal of Urban History* 5, 308–339.

Tarr, J.A., McCurley, J., McMichael, F.C. and Yosie, T. 1984. Water and Wastes: A Retrospective Assessment of Wastewater Technology in the United States, 1800–1932. *Technology and Culture* 25, 226–263. doi:10.2307/3104713

Thye, Y.P., Templeton, M.R. and Ali, M. 2011. A Critical Review of Technologies for Pit Latrine Emptying in Developing Countries. *Critical Reviews in Environmental Science and Technology* 41, 1793–1819. doi:10.1080/10643389.2010.48 1593

Travers, K., Khosla, P. and Dhar, S. 2011. *Gender and Essential Services in Low-Income Countries*. Women in Cities International and Jagori, Montreal and New Delhi.

UNICEF and WHO. 2015. *Progress on Sanitation and Drinking Water – 2015 Update and MDG Assessment*. WHO Press, Geneva.

United Nations. 2015. *Sustainable Development Goals* [WWW Document]. Sustainable Development Knowledge Platform, Department of Economic and Social Affairs. https://sustainabledevelopment.un.org/?menu=1300.

United Nations General Assembly. 2010. *The Human Right to Water and Sanitation* (Resolution adopted by the General Assembly No. A/RES/64/292). United Nations, New York.

WHO and UNICEF. 2017. *Sanitation* [WWW Document]. JMO. https://washdata. org/monitoring/sanitation.

7 Drainage

Introduction

Drainage is the most mundane of infrastructures yet it holds the potential to profoundly alter relationships between cities and nature. Draining away rainwater that falls on roofs, streets and hard surfaces in cities is an essential infrastructure service. Drains are amongst the most ancient features of cities, designed to prevent surface water flooding and unplanned, unhygienic standing water in urban environments. How this water is managed and discharged to the environment has important implications for urban sustainability. Surface water is a major source of pollution in many cities. Sustainable approaches to surface water management can bring wider benefits, including urban cooling, improved biodiversity and habitat, and social amenity.

Sustainable drainage represents a fundamentally different philosophy of water management in cities. For most of the twentieth century urban drainage systems were dominated by heavy engineering, with their primary function being to remove water from the urban environment as quickly as possible to prevent flooding. These fast conveyance systems typically require large networks of underground pipes and drains which are either combined with urban wastewater or an entirely separate system of pipes that discharge directly into the local environment (Butler and Davies, 2010). The dominance of the fast conveyance approach to urban drainage has contributed to the conversion of urban waterways to drains and culverts. Covering over streams and rivers has also provided space for urban development, particularly for road construction. The loss of urban rivers and associated biological diversity has become an issue of increasing concern to urban environmentalists.

Since the 1970s urban drainage systems have been under scrutiny as a source of pollution (Brown et al., 2009). Surface water runoff discharged directly into the environment causes pollution from oil and grease, litter, grit, heavy metals, pesticides, fertilisers and other contaminants that are washed or dumped into drains across the city. Combined sewer overflows (CSOs) have become a significant source of pollution, discharging dilute, untreated sewage into urban rivers. Originally intended to occur infrequently, CSOs have become common environmental pollution events as urban surfaces have been built upon and paved over, and population growth has increased wastewater volumes.

In response to concerns about the environmental impacts of urban drainage systems, engineers, ecologists and urban designers have devised new approaches to managing surface water. Water sensitive urban design (WSUD) in Australia, low impact development (LID) in the US and sustainable drainage systems (SuDS) in the UK have emerged as sustainable approaches and techniques for urban drainage that mimic natural hydrological systems (Figure 7.1). Rather than prioritising fast conveyance of surface water away from buildings, streets and urban spaces, WSUD, LID and SuDS attempt to increase local infiltration and storage of water and attenuate flows of stormwater, minimising peak flows, preventing surface water flooding and reducing pollution. In recent decades, approaches to the design of urban drainage systems have begun a transformation towards working with natural hydrological flows and ecological systems after more than a century of controlling and excluding surface water from cities. Sustainable urban drainage becomes an integrated element of green infrastructure in cities, providing multifunctional benefits.

This chapter introduces the principles and techniques of sustainable drainage and its wider benefits to cities. Urban drainage is then discussed from the point of view of the five frameworks of sustainability. Urban drainage is highly significant for the sustainability of cities, not simply as a solution to the essential problem of how to manage water that falls on urban surfaces, but as an indication of alternative ways for cities to relate to nature and landscapes.

Rainfall, runoff and water quality

The volume and speed of runoff during and after a storm event depends upon the permeability of the surface, soil and groundwater conditions, the topography of a site or catchment, and the duration and intensity of the rainfall (Figure 7.1). Pre-development 'greenfield' conditions tend to allow for water to infiltrate the soil, reducing runoff from the site. Urbanisation increases hard surfaces and reduces opportunities for infiltration, increasing the volume and speed of runoff. Urbanisation also introduces sources of contaminants such as oil and grease leaking from cars, grit and salt from roads, and fertilisers and pesticides from lawns and gardens. Managing the quantity and quality of urban runoff is a major task for engineers and planners and an important element of integrated urban water management. Sustainable approaches to drainage aim to reduce the overall volume and improve the quality of urban runoff to the environment or sewers, mimicking pre-development conditions within the urban environment.

The volume of runoff produced during and after a storm event is typically presented as a hydrograph (Figure 7.2). This is a graph which shows

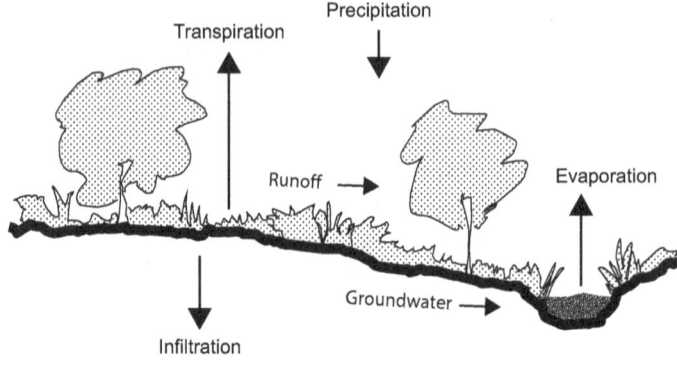

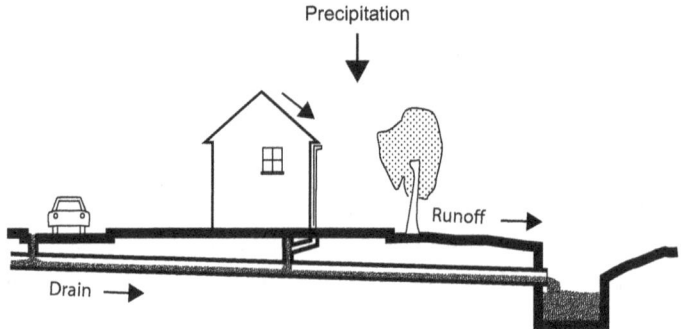

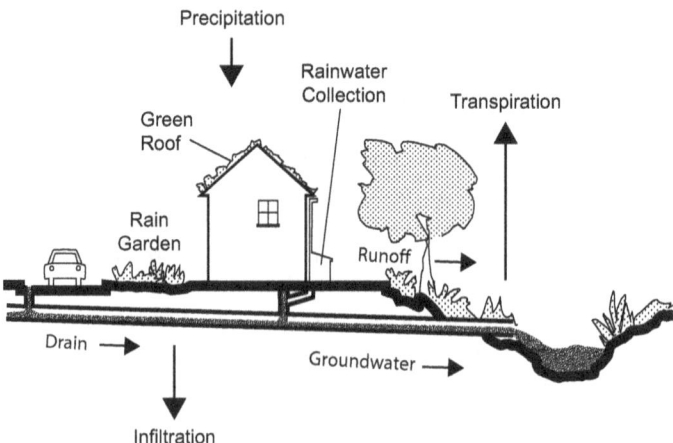

Figure 7.1 Natural, conventional and sustainable drainage

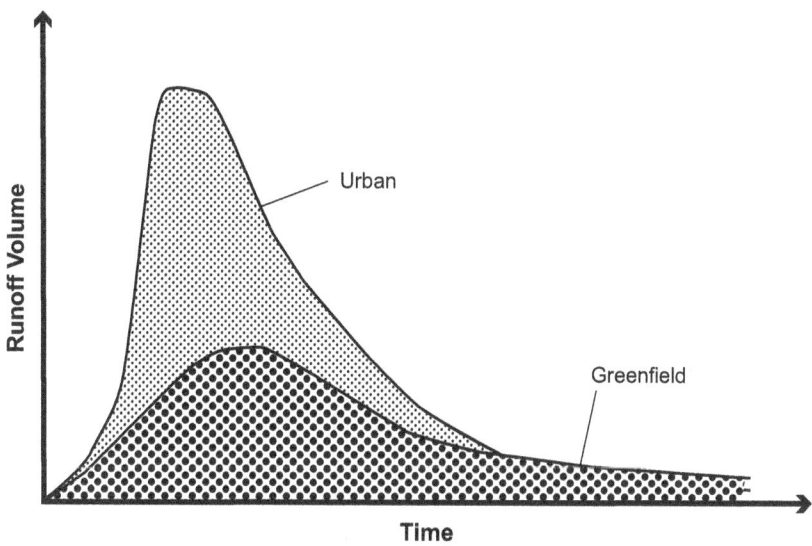

Figure 7.2 Urban stormwater hydrographs: pre- and post-development

the volumetric flow rate over time at a particular point in a river, drain or floodplain. In urbanised catchments with low permeability, water will run off quickly, with a large peak in flow rate soon after the storm begins that declines quickly once the storm passes. For the same storm, a greenfield site with higher infiltration and storage will take longer to reach peak runoff and the peak will last for longer, but it will be lower than for the urbanised catchment. The total volume of runoff is the area under the hydrograph curve, and this is lower for the greenfield site because of greater infiltration and storage. Sustainable drainage aims to achieve greenfield runoff conditions in urbanised catchments, increasing infiltration, evapotranspiration and storage on-site to slow and reduce surface water runoff (Figure 7.1).

Reducing peak runoff has several benefits to urban infrastructure and environments. Fast-flowing runoff discharged into local streams and channels causes erosion and scouring of banks and flood zones, disrupting habitats and ecosystems. Where runoff discharges to piped drainage networks, drains must be designed to be large enough to convey the peak flow, increasing the scale and cost of drainage infrastructure. High volumes of runoff accumulating quickly in urban areas can exceed the capacity of drainage systems, causing surface water flooding. Reducing the total volume and attenuating the peak flow of urban runoff provides relief to receiving waters and engineered drainage networks, and reduces the risk of surface water flooding.

Designing urban drainage systems requires anticipating the likely rainfall events that will be experienced in a city. Cities which experience monsoon rains must design drainage networks to deal with very large rainfall events, while cities with temperate climates typically design for rainfall of less intensity but of a longer duration. Design guidance for urban drainage and flood protection typically refers to the probability or 'return period' of rainfall or flood events. A 1:100 year rainfall event has a 1% probability of occurring in any single year. A 1:1 year event is likely to occur every year. In deciding what probability event to design for, planners and engineers trade off economic and social costs and impacts. Designing for a 1:1,000 year event would provide a very high level of protection against flooding during very rare storm events, but would typically require a high level of investment and large-scale infrastructure. The costs of investing in and maintaining such a high level of protection usually outweigh the benefits but may be justified in protecting areas of high population and economic value. The Thames Barrier (Figure 7.3) is designed to protect central London from a 1:1,000 year storm surge from the North Sea due to the catastrophic nature of such a flood for the whole city, but drainage standards in London provide protection from on-site surface water flooding from a 1:30 year rainfall event (DEFRA, 2015).

Specification for levels of protection and suggested management measures vary in different jurisdictions, with increasingly detailed standards

Figure 7.3 The Thames Barrier

set at national, provincial and local levels. Regulations at the national and regional levels typically set out general principles and guidance, which are then translated into specific local requirements set by individual planning authorities. Local planners may specify requirements for particular zones based on local sewer capacity, environmental conditions and site characteristics. Local manuals and standards reflect local hydrological and environmental conditions and recommendations for the most effective drainage infrastructure.

Climate change is likely to change rainfall patterns in many places. Translating climate change forecasts into local rainfall events remains highly uncertain, and climate change allowances for drainage design may be as simple as requiring a factor of safety, for instance a 20% increase in rainfall, in system capacity and attenuation capability (Woods Ballard et al., 2015). Planners may also require additional capacity to be designed into the scheme to account for 'urban creep', allowing for reduced permeability of the site over time due to changes in land surface.

Stormwater runoff can become polluted by materials that are commonly found on urban surfaces such as roads, driveways, roofs and gardens, and by spillage or dumping of chemicals and waste (Brinkmann, 1985). Runoff from roads and car parks can be polluted with sand and grit, salt and de-icing chemicals, oil and grease from vehicle leaks, and heavy metals and particulates from wear and tear on tyres. Fertilisers, pesticides and sediments can pollute runoff from gardens, particularly lawns and playing fields. Atmospheric deposition of particulates, hydrocarbons and metals onto roofs, roads and other surfaces can contribute to water pollution as they are washed off surfaces during storm events. Washing vehicles, pavements, roads and outdoor equipment such as rubbish bins contaminates surface water with detergents, bacteria, viruses, particulates and whatever else is the target of cleaning. General litter, animal faeces, deliberate dumping of waste and misconnection of sewers are also sources of multiple forms of pollution of surface water. Surface water runoff washes away whatever contaminants are found on open surfaces in cities and is further polluted by deliberate dumping of waste into drains.

The concentration of pollutants in stormwater runoff is not uniform throughout a storm event. The 'first flush' rainfall event after a dry period is likely to have the highest concentration of pollutants as contaminants built up on surfaces or in drains are washed off or re-entrained in water flows in pipes and channels (Bertrand-Krajewski et al., 1998). Concentration of many contaminants typically peaks early in the storm event, declining rapidly as the pollutants are washed from the surface. Managing the first flush of contaminants is therefore an important goal for reducing pollution from surface water runoff. This can be achieved by diverting the first flush away from the receiving environment towards storage or treatment (Kayhanian and Stenstrom, 2005). Sediment contamination of runoff may continue to increase as a result of erosion in the catchment, from roadsides, building sites and in channels subject to high flow volumes and velocities. The quality of surface water and the risks of contamination are important considerations in designing drainage systems that aim to increase groundwater infiltration as the means of reducing runoff. Infiltration of contaminated surface water can contribute to pollution of groundwater, with long-term environmental consequences due to the difficulty of remediation and containment.

Combined sewer overflows (CSOs) are a specific challenge for reducing pollution in cities in North America and Europe. Combined sewers were built in the nineteenth and twentieth centuries to transport both wastewater and surface water runoff. In order to avoid flooding homes and streets with dilute wastewater, these networks are designed to overflow during high rainfall events (Figure 7.4). Urbanisation, particularly paving over or developing previously permeable surfaces such as parks and gardens, has increased urban runoff. This means that sewers fill up and overflow during relatively frequent storm events. In London CSOs occur approximately 50 times per year on average. Discharging dilute sewage into local rivers is a significant source of pollution. In the US, Consent Decree Orders from the Environmental Protection Agency to individual cities require action to reduce CSOs and improve urban water quality to meet the Clean Water Act. In the European Union the Urban Wastewater Directive and the Water Framework Directive include requirements to improve the ecological status of rivers and reduce urban pollution, including reducing CSOs. Conventional infrastructure responses to CSOs are based on interceptor tunnels or other storage to collect and treat overflows before they pollute rivers. Green infrastructure solutions aim to reduce runoff into the sewers, preventing overflows from occurring. The applicability of different solutions to CSOs is determined by a range of technical, environmental and political factors, including soil conditions, urban form, infrastructure ownership and regulation, access to finance and the severity of pollution to be mitigated.

Sustainable drainage

Sustainable drainage approaches utilise techniques and technologies to increase infiltration and storage, reduce surface water runoff, improve water quality, and mimic natural flow patterns (Ahiablame et al., 2012; Coutts et al., 2013; Ellis, 2013; Mitchell, 2006). Technologies can include wetlands, ponds and ecological systems, as well as innovative 'hard' drainage technologies for slowing runoff, storing water and controlling flows. However, SuDS, LID and WSUD are systems of drainage, not isolated techniques or technologies. Individual elements may perform multiple functions, but sustainable drainage is an integrated approach to achieving overall goals of reduced runoff, improved water quality, amenity and biodiversity. Most schemes involve multiple elements acting at different scales that combine to meet overall objectives for the site and its contribution to broader goals of urban sustainability.

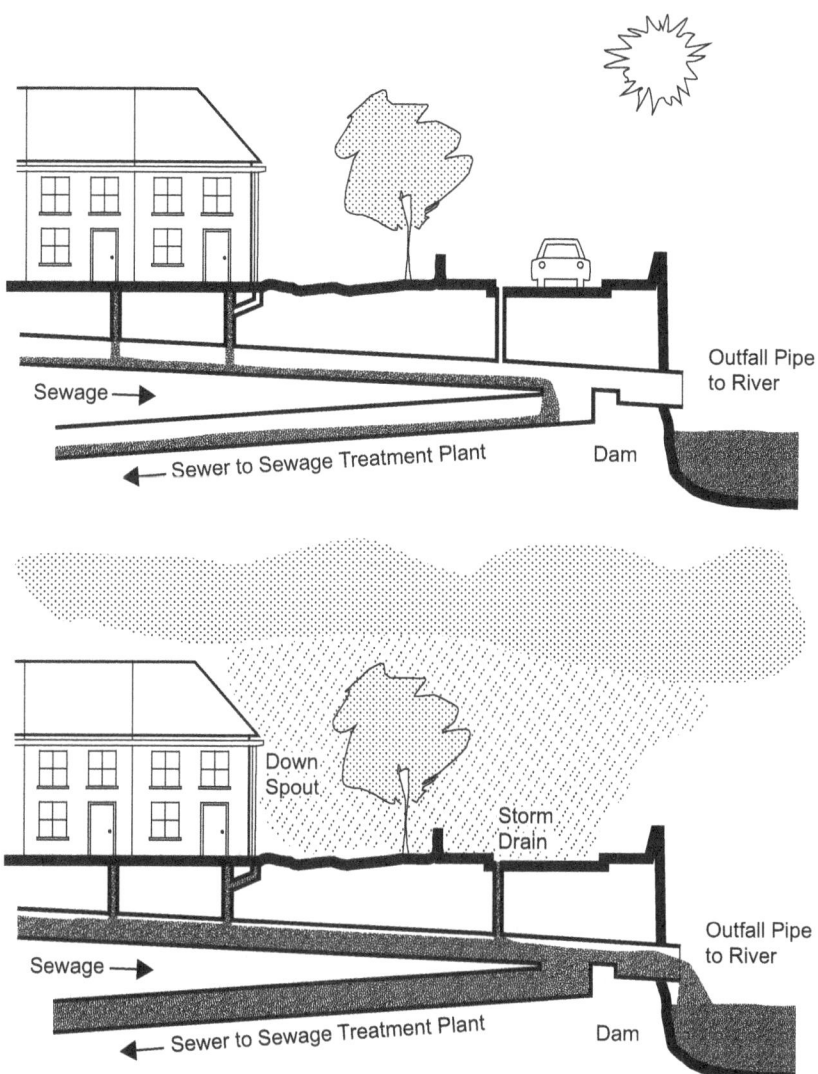

Figure 7.4 Combined sewer overflows

Reducing pollution of groundwater and receiving waters from contaminated surface water requires an integrated approach across the urban catchment. Removing sources of pollution such as littering, excessive application of fertilisers and pesticides, and point source leaks is an important first step, requiring community awareness and appropriate regulation and enforcement. Containing and treating contaminated runoff on-site, particularly the first flush, reduces environmental pollution. Sustainable drainage techniques are also able to provide treatment of surface water to remove sediments, nutrients, oil and grease, and heavy metals through physical separation, settling and biological uptake (Ahiablame et al., 2012; Loperfido et al., 2014). Managing drainage systems to ensure that pollutants are stabilised and removed is important to avoid re-entrainment or release back into the environment over time (Ellis, 2013).

In attempting to mimic natural hydrological systems, sustainable drainage design follows a hierarchy of scales from source, to site, to region. Source control measures such as reducing impermeable surfaces and installing green roofs, rainwater harvesting and permeable pavement aim to prevent surface water runoff. Site control measures such as swales and rain gardens are designed to contain runoff locally. Regional control measures such as balancing ponds, lakes and wetlands receive surface water runoff from local sites, providing a buffer to receiving environments. Conveyance between elements of a sustainable drainage scheme, within the site and across scales, is also an important design consideration which can contribute to infiltration, storage and treatment. At all stages water is infiltrated and evaporated and may be discharged to the local environment following treatment.

Rainwater harvesting is a source control measure which fulfils the dual function of storing surface water from roof runoff and providing an alternative source of water for supply (Campisano et al., 2017). Water supply and surface water management objectives can be balanced in design and management of rainwater harvesting systems (Melville-Shreeve et al., 2016; Mugume et al., 2016). For water supply, the objective is to have sufficient water stored in the tanks to meet demand, while for surface water management the objective is to have sufficient space in the tank to contain roof runoff. These competing objectives can be managed by optimising tank sizing according to rainfall and demand patterns, and through operational strategies. For instance, tanks can be managed so that long-term storage is maintained below half-full to provide room for attenuating storm events, while still providing water for supply. Smart rainwater harvesting systems enable remote control of tank levels to achieve design and operational objectives. Systems are under development that allow for tanks to be emptied in advance of forecast rainfall events, so that storage is available when it is needed.

Green roofs store small rainfall events and attenuate larger storms (Czemiel Berndtsson, 2010; Teemusk and Mander, 2007). Rainfall is stored in the green roof substrate and runoff is slowed down by the roughness of the green surface compared to conventional roofing materials. Green roofs increase evapotranspiration, reduce runoff and provide temperature regulation of the building and surroundings (Figure 7.5). Green roofs can also contribute to improved biodiversity, provide habitat for insects and birds, and filter pollution deposited on the roof surface.

Permeable pavements reduce surface water runoff by improving infiltration to the underlying ground and providing storage and attenuation of

Figure 7.5 Green roof with solar cells and food growing in Portland, Oregon

water within the paving system itself. Permeable pavers have high porosity, providing storage, and allowing water to infiltrate to underlying aggregate or subsurface drainage for further conveyance, treatment and storage. However, they are prone to blockage if not maintained, as grit and plants can fill pores and reduce permeability (Bean et al., 2007).

Disconnecting downpipes from buildings and redirecting flows to soakaways can reduce runoff by increasing infiltration. This simple measure is appropriate only in soils with high permeability and on sites with low risk of contamination of groundwater. Downpipes may also be directed to rain gardens, which can control runoff at source or at site level, receiving runoff from roads or other drainage elements.

Rain gardens are examples of bioretention systems, integrating physical storage, evapotranspiration, infiltration and biological treatment of runoff (Davis et al., 2009). Rain gardens are designed with deep substrate to store and filter water (Figure 7.6). Plants are selected to draw on the subsurface water during dry periods and to withstand inundation during rainfall events. Treatment performed by rain gardens can include settling and filtering of particulates, uptake of nutrients, separation of oil and grease, and stabilisation of heavy metals (Davis et al., 2003, 2006).

At the site level, conveyance and infiltration can be achieved through swales which slow the flow of water across the site, but provide a pathway for controlled flow between drainage system elements. Swales may be planted to provide treatment and evapotranspiration. Bioretention basins

Figure 7.6 Rain garden receiving runoff from pavement and road in Portland, Oregon

and ponds provide storage, treatment and biodiversity and amenity values, and can be designed to fill during storm events and release water slowly to reduce peak runoff. Site features such as football pitches, basketball courts and playgrounds can also provide storage and attenuation through controlled inundation and drainage during and after storms. Detention basins, wetland systems, river restoration, and parks and open spaces provide measure for regional control of surface water, allowing storage, infiltration, treatment and evapotranspiration prior to release to the environment. Trees in urban catchments also serve to improve drainage, slowing surface water runoff, improving infiltration and evapotranspiration.

Sustainable drainage can deliver multiple benefits to cities as part of blue and green infrastructure networks (Kambites and Owen, 2006; Voskamp and Van de Ven, 2015). In addition to meeting objectives for surface water management, green infrastructure approaches can improve local amenity, including place-making and community development, provide spaces for play and recreation, improve habitat for plants and wildlife, reduce urban heat island effect through shading and cooling from evapotranspiration, and improve health and wellbeing through improved access to green space and a greener environment (Dover, 2015). Sustainable design of drainage should therefore engage with local communities and stakeholders to identify local needs and opportunities, in addition to the design requirements for water quality and quantity.

Distributed technologies provide new opportunities for monitoring and managing drainage systems. Sensor networks can monitor water levels, flow rates and water quality, and cameras allow remote monitoring of surface water levels and the overall performance of the systems. This provides opportunities for centralised management and monitoring of localised drainage systems, and it allows for optimal management of competing objectives such as rainwater harvesting for supply and storage, and multifunctional use of open space for recreation and stormwater attenuation.

Urban river restoration reflects a changing approach to managing surface water in cities, which acknowledges the importance of urban aquatic ecosystems. In many cases this has involved the removal of concrete channels and returning natural meanders and vegetation to small streams. Improving the quality of water discharged into these water courses is also important, through better control of industrial pollution as well as improved quality of stormwater discharge and reduced frequency of combined sewer overflows. In some cities river restoration has resulted in relatively modest, gradual improvement in the local environment as concrete drains are returned to functioning ecosystems. In other cases it has involved dramatic transformation of the built environment and infrastructure, such as in Seoul where the restoration of the Cheonggyecheon River involved the removal of a major motorway (Kang and Cervero, 2009).

Framing drainage

Urban drainage is an essential element of the built environment which has a major impact on urban form and design. Conventional urban design and engineering approaches keep surface water hidden in all but the most controlled circumstances. Drains are buried and open channels are hidden behind walls. Sustainable drainage seeks to fundamentally change the role of water in the urban environment from a problem to be contained, hidden and removed as quickly as possible, to reveal a resource with multiple benefits and a lively element of urban design. Sustainable drainage aims to mimic natural hydrology, making space for water as part of the urban environment. Moving beyond the usual concerns about reducing pollution and resource use, sustainable drainage is significant in breaking down cultural and conceptual distinctions between 'nature' and 'the city', emphasising the value of recreating natural hydrological systems in urban environments (Jones and Macdonald, 2007; Karvonen, 2010).

Proponents of sustainable drainage, water sensitive urban design, low-impact development, green infrastructure and associated approaches make a strong case for the unsustainability of existing drainage strategies and the benefits of their 'new paradigm' and new techniques. However, urban drainage does not exist in isolation. While sustainable drainage is taken as indicative of a new paradigm of water management, it exists as part of debates and discussions about water and environmental policy. As such, drainage and how to implement more sustainable approaches have been subject to analysis under frameworks of urban water sustainability.

Sustainable development

The global emergence of sustainable development provided a foundation for justification of a new approach to drainage promoted in the 1990s in different parts of the world. Early proponents of sustainable drainage approaches referred to the needs of future generations, ecological limits to development and working within local environmental conditions as justification for new techniques within a broader shift in philosophy of urban water management (Butler and Parkinson, 1997; Cettner et al., 2014; Hedgcock and Mouritz, 1993; Niemczynowicz, 1999). Local Agenda 21 also provided a rationale to focus on local government action on drainage, a long-standing municipal service subject to local town planning and development control. Traditional techniques of fast conveyance were characterised as unsustainable, requiring a 'new paradigm' of water management in cities. In 1997 Butler and Parkinson proposed that sustainable urban drainage should

- maintain an effective public health barrier and provide sufficient protection from flooding;
- avoid local and more distant pollution of the environment (air, land, and water);
- minimise the utilisation of natural resources (e.g. water, energy, materials); and
- be operable in the long term and adaptable to future requirements.

In development discourse drainage is often conflated with either sanitation or flood protection, which misses the specific role that drainage infrastructure plays and opportunities for green infrastructure to transform urban environments. Drainage is not mentioned specifically in the SDGs, although it is central to achieving targets for protecting aquatic ecosystems, reducing pollution and integrating water management in Goal 6 and reducing water-related disasters in Goal 11.

The systemic consequences of changing permeability and drainage routes are frequently overlooked in urbanisation processes, leading to increased risk of surface water flooding, pollution and ecological degradation as elements of natural or traditional drainage systems are built over. Informal settlements are notoriously located in floodplains, wetlands and hillsides, contributing to runoff as well as being vulnerable to flooding. However, formal urban development processes also often involve reclamation of wetlands, building on floodplains, reduced surface permeability and redirecting natural drainage flows, contributing to pollution and flood risk. The city of Bangalore in India has experienced an increase in flooding in recent years, partly as a result of infilling of traditional tanks and drainage channels to create new land for housing and commercial development (Ranganathan, 2015; Sundaresan et al., 2017).

Drainage is a core element of urban upgrading programmes. Parkinson et al. report that runoff from slums is typically lower than would be expected from urban areas due to high permeability of unpaved surfaces, evapotranspiration from remnant vegetation, storage in depressions and disjointed and poorly connected drainage channels (Parkinson et al., 2007). While separate sewers are generally preferred for sustainable surface water management, in slum upgrade projects, just as in nineteenth-century London, combined sewers may be preferable as a low-cost, compact solution to public health crises in complex, dense urban environments. Rainwater harvesting can be incorporated in urban improvements as a supply of water for household use and to reduce runoff (Mugume et al., 2016; Parkinson et al., 2007). Upgrading also requires consideration of the impact of other urban services, such as solid waste and roads, on drainage. The absence of solid waste management infrastructure can increase blockage of drains and pollution of local rivers and wetlands. Paving roads can reduce inflow of silt to drains but increase runoff due to reduced permeability. Draining roads is important to maintain access during high rainfall events.

The Slum Networking approach developed by Himanshu Parikh in India highlights the connections between slums as part of the city, identifying opportunities to improve overall environmental and social conditions through the provision of infrastructure (Parikh and Parikh, 2009). Natural drainage paths provide the basis for drainage and other infrastructure networks, making best use of gravity and topography for road construction as well as drainage. The approach also emphasises the importance of planting and infiltration to reduce surface water runoff and improve environmental quality. Urban upgrading therefore provides opportunities for source control and sustainable drainage measures, rather than replicating fast conveyance systems that have dominated cities in the Global North since the nineteenth century.

Ecological modernisation

Sustainable drainage is not obviously compatible with ecological modernisation policy. It requires major reform of institutions and design philosophies, there are no existing market-based mechanisms to drive innovation, and it is not directly focussed on resource or economic efficiency. In order to demonstrate the relevance of sustainable drainage within a dominant ecological modernisation and neoliberal policy discourse, proponents have emphasised ecosystem services and triple bottom line benefits. This broadens drainage project and policy evaluation from purely economic valuation methods, while reformulating a programme that is grounded in environmental improvements into narrower economic terms in order to facilitate policy uptake.

Ecosystem services has become the preferred language for demonstrating the importance of nature to the human economy, as the means for ensuring environmental protection and restoration. A famous paper by Costanza et al. (1997) estimated the economic value of services provided by nature to be US$33 trillion per year. While the final figure and methods of valuation used in this study have been criticised, the concepts of ecosystem services and natural capital have gained traction in popular and policy discourse (Braat and de Groot, 2012). Putting a value on ecosystem services demonstrates the benefits to society of investing in environmental conservation and green infrastructure, including benefits to health, resource provision, recreation and pollution reduction.

Analysis of the ecosystem services provided by sustainable drainage and green infrastructure are intended to demonstrate the wider benefits of new approaches compared to the simple drainage benefits of conventional drainage infrastructure. This can be used to justify higher costs and risks associated with sustainable drainage. However, a core challenge for ecosystem services analysis is that the benefits are not necessarily accrued to the owner of the infrastructure who incurs the higher costs and risks. For instance, a local highways authority may not receive any economic or other benefits from improved habitat or recreational opportunities for a sustainable drainage scheme that may cost more and be more difficult to maintain compared to conventional drainage. The Philadelphia Green City, Clean Waters programme (Box 7.1) required improved communication between government departments to enable delivery of a green infrastructure strategy for solving CSOs, supported by a triple bottom line analysis showing a wide range of benefits beyond water management (Fitzgerald and Laufer, 2017; PWD, 2009).

Box 7.1   Philadelphia Green City, Clean Waters

In 2011 the City of Philadelphia reached agreement with the US EPA on a 25-year strategy to halve the volume of CSOs by capturing 85% of the volume of runoff to combined sewers using green infrastructure techniques (PWD, 2011). This was the first such agreement in which CSO pollution will be addressed entirely using green infrastructure. The 'Green City, Clean Waters' programme involves investment of US$2.4 billion by the Philadelphia Water Department over the length of the programme, with total investment including partner organisations and the private sector anticipated to be in excess of US$3 billion. In recent decades Philadelphia has experienced a loss of industry, employment and population. The decision to pursue a green infrastructure solution to CSOs was informed by a triple bottom line analysis which showed a greater range of benefits at lower upfront capital costs compared to conventional solutions based on interceptor tunnels. The green infrastructure strategy reduces the financial burden on the city through lower capital costs and provides more local employment and economic opportunities than a conventional tunnel project. It is also anticipated to deliver wider benefits including improved air quality, recreational opportunities and urban cooling. Implementing the programme has required innovation in governance and financing (PWD, 2009). The programme is led by the Water Department but impacts on the work of a range of departments in the city government, particularly the Parks and Recreation and the Streets departments, in a context of financially constrained local government in a city that has experienced economic decline (Fitzgerald and Laufer, 2017). Delivering the plan has therefore required communication and integration of different priorities and plans across local government departments, addressing complexities such as the impact of green infrastructure on traffic flows and disability access in the Streets Department, and park users' concerns about the appearance of rain gardens. Since 2010 non-residential properties have been charged a stormwater fee based on impervious area and in 2014 a charge was introduced for residential properties based on parcel size. Private landholders are encouraged to invest in source control and green infrastructure through rebates on stormwater charges, loans and grants, as well as through planning requirements on new development and redevelopment. In order to encourage implementation of source control measures on private land at sufficient scale and reasonable cost, third-party aggregator organisations act as intermediaries between the city and landholders and small businesses in managing grants, rebates and contracts for project delivery (Valderrama et al., 2013). Approximately US$2 million is spent on public outreach and education, including education programmes delivered by Parks and Recreation. The first five-year target to reduce stormwater runoff and CSOs by 1.5 billion gallons (5.7 billion litres) was exceeded in 2016, through the creation of 838 greened acres (339 hectares) across 441 green infrastructure sites (PWD, 2016).

Framing sustainable drainage within an ecological modernist approach has contributed to the development of innovative mechanisms for financing. Drainage has traditionally been a public service requiring high capital investment by local government authorities. In the UK some elements of the drainage network have been privatised as part of the sewerage infrastructure, but drainage networks have typically remained public assets designed and operated to service public and private land and buildings. Financing drainage investment and maintenance may be incorporated in water charges, or paid from central budgets. The implementation of specific stormwater charging in cities such as Portland, Washington, DC, and Philadelphia has provided funding and incentives for various grants, rebates and trading schemes to encourage landowners to reduce runoff to sewers (Valderrama et al., 2013). China's Sponge City programme provides a national-scale trial in the potential for public-private partnerships in financing drainage infrastructure (Box 7.2).

Box 7.2   China's Sponge City programme

In July 2012 a heavy rainstorm caused flash flooding in Beijing. More than 1.6 million people were affected and 79 people were killed. Damages were estimated to be in excess of US$2 billion (Chen et al., 2014). While tragic and costly, this event was not unusual in China, where rapid urbanisation has led to frequent, severe surface water flooding. The Beijing event contributed to a significant change in policy in China to address the risk of urban flooding. The Sponge City programme was launched in December 2014 as a joint initiative between the Ministry of Finance and the Ministry of Housing and Urban-Rural Development. The programme provided central government funding for cities to initiate urban drainage programmes with a particular emphasis on sustainable drainage, green infrastructure and low-impact development approaches. Around 200 cities applied to the programme, and in 2015 16 cities were awarded funding as Sponge City Pilots, followed by a further 14 cities added in 2016. While the focus is on green infrastructure measures, pilot city programmes also include wastewater treatment plants and interceptor tunnels for stormwater management. The pilot projects were awarded between US$48–80 million each, for a total of US$6 billion over three years. The central government funding is not intended to support the full cost of construction, but is to be matched by local government funds and private sector investment through public-private partnerships (PPP). The central government funding is intended to stimulate further investment and employment as an economic stimulus measure as well as improve urban resilience and water management. The pilots provide an opportunity to address challenges of governance and financing sustainable urban water infrastructure, as well as building technical capacity to design and deliver low-impact development measures (Jiang et al., 2017). Translating central government ambitions for sustainable drainage into projects delivered on the ground faces considerable challenges in terms of interdisciplinary technical expertise, design and planning. The long-term viability of the programme remains to be confirmed. Key uncertainties beyond the pilot cities include the level of government funding and the feasibility of PPP in a sector without well-established, profitable business models. Private sector interest has been stronger in wastewater and conventional infrastructure projects than for drainage and green infrastructure, and public sector funding remains core to delivering the public benefits of reducing vulnerability to flooding and improving urban environments.

Socio-technical systems

The technical promise of sustainable drainage is unlikely to be achieved within governance and regulatory institutions that have evolved with the fast-conveyance model of surface water management and engineering. Drainage institutions, like stormwater drains themselves, have been largely invisible in cities, part of mundane building regulations and municipal works departments. Technical innovation, modelling and evaluation of sustainable drainage techniques are necessary but not sufficient for city-scale adoption. Delivering multiple benefits of decentralised infrastructure requires new forms of governance, ownership, financing and regulation.

Analysing drainage networks as socio-technical systems shows how technical standards and norms are embedded within institutions, professions and knowledge networks. Drains are literally 'locked-in' to cities in the Global North, buried under roads and buildings. They are also 'locked-in' to wider practices of urban development, through building codes, professional expertise, planning regulations and institutional structures.

Rebekah Brown and colleagues have studied drivers and barriers to achieving sustainable urban water management in Australia, drawing on socio-technical transitions theories (Brown et al., 2006; Brown and Farrelly, 2009; Rijke et al., 2013). A case study of the transition to sustainable urban water management in the Cooks River catchment in Sydney began by focussing on stormwater. Bos and Brown (2012) map the evolution of a programme of action from an initiative driven by a handful of 'champions' who were driven to improve stormwater management, to an integrated programme across six local governments, state government, industry and community stakeholders. The Cooks River case shows the important of experimentation in achieving change in urban water management. Technical experimentation in drainage is relatively widespread, but in order to scale up the approach and achieve lasting changes in practice new forms of governance and management are also required. Experimentation in governance involves the formation of new institutions to address the complexity of integrated and sustainable management (Brown et al., 2009; Brown et al., 2013; Farrelly and Brown, 2011).

Experimentation and learning are also evident in Philadelphia's implementation of the Green City, Clean Waters programme (Box 7.1). Fitzgerald and Laufer (2017) document the process of change within the City of Philadelphia as the programme was conceived and in the early stages of delivery. The decision to pursue a strategy for reducing CSOs based entirely on green infrastructure emerged from a history of innovation in water management governance within the Philadelphia Water Department, but its implementation required greater communication and integration with other departments, including the Streets and the Parks and Recreation departments. Fitzgerald and Laufer note that city officials and professionals resisted the characterisation of innovation as 'experimentation', preferring to emphasise the deliberate and expert judgement involved in piloting and adapting different measures as the plan was implemented. The reluctance of water and other professionals in Philadelphia to identify governance reform and innovation as 'experimentation' accords with Farrelly and Brown's (2011) finding that risk aversion remains a barrier to the transition to sustainable water management.

Transitions theory helps to explain why technical and governance innovations succeed or fail in sustainable drainage as in other sectors of infrastructure and socio-technical systems. The multi-level perspective highlights the need for alignment of regime- and landscape-level institutions and cultures, as well as encouraging niche-level innovations. Farrelly and Brown (2011) identify misalignment between different levels of government and professional cultures as a factor constraining the transition to sustainable water urban management in Australia. Philadelphia's success so far demonstrates the value of multi-level innovation and reform. At the 'landscape' level, the plan was confirmed by the US EPA in the Consent Decree, an outcome of innovation and negotiation. This further strengthens support within the 'regime' level of the City, allowing for ongoing innovation in technical, social, regulatory and financial mechanisms at the 'niche' level.

Urban political ecology

Urban runoff is a quintessential socio-environmental issue. Rain falling from the sky is a natural process, urban rivers, streams and wetlands are remnants of pre-urban landscapes, and the control of surface water has been a primary engineering and administrative task for urban societies and authorities since ancient times. Urban runoff is natural, technical, social and political. Dominant forms of drainage infrastructure aim to control flows and remove water from cities as quickly as possible, attempting to enforce the separation of nature and cities. Environmental pollution, altered geomorphology of rivers and lakes, and flooding are indicators of the fallacy of this ideal. Alternatives based on mimicking natural hydrology using green infrastructure and distributed technologies promise recognition of nature and water as part of, rather than separate from, urban life.

Andrew Karvonen's book *The Politics of Urban Runoff* demonstrates how trends in drainage reflect broader social, political and environmental processes (Karvonen, 2011). His studies of Austin and Seattle in the US trace the history of drainage infrastructure and politics, moving from the 'Promethean' era of taming and controlling water and landscapes to serve urban development, towards efforts to prevent environmental pollution, restore rivers and implement innovative natural drainage schemes. Drainage infrastructure is tied to wider urban politics, with unequal costs and benefits. Historical schemes to improve drainage and reduce pollution in some parts of Seattle displaced water flows and waste to poorer neighbourhoods. More recently river restoration and sustainable drainage have been used to placate local residents objecting to urban development. Decisions about drainage reflect changing environmental politics but remain subject to underlying political forces (Karvonen, 2010).

Urban drainage is subject to political discourse just as any other infrastructure or feature of public life. In Los Angeles, Cousins (2017) identifies four distinct discourses in debates about drainage – market sceptic, managerialist, market technocrat, and the regulatory and administrative technocrat. These discourses shape preferred institutional and policy frameworks for drainage, which in turn influence the choice of technologies and strategies for managing surface water. In Pittsburgh, Finewood (2016) shows how green infrastructure was co-opted by a dominant 'grey epistemology', supporting existing structures of knowledge, power and decision-making. Green infrastructure in Pittsburgh came to be accepted by drainage engineers and managers in the municipal authority only to the extent that it did not displace the basic function of conventional drainage infrastructure or undermine their role as experts driving decision-making.

Urban politics are also at play in processes of urbanisation in the Global South, which have undermined natural and traditional drainage structures, increasing pollution and flood risk. In Bangalore, Ranganathan (2015) describes the transition of traditional tanks and channels that served both drainage and water supply functions into sanitary sewers, used to transport wastewater, during the colonial era. More recent growth in the city based on its role in the global IT sector has driven conversion of tanks, channels and wetlands into property developments including housing, commercial offices, golf courses and a sports stadium. Global capital has become a driver for infilling natural and traditional water courses, thereby exacerbating pollution and flood risk. Urban drainage is therefore entangled in global networks of finance and capital, as well as local politics and planning.

Radical ecology

Making space for water, mimicking natural hydrological flows, creating habitat and reconnecting with natural flows and cycles represent an accommodation of water in cities and adaptation of urban settlements to the patterns of nature rather that constraining and controlling water and nature to fit human designs. Rather than dominating and controlling nature through pipes and engineered infrastructure, sustainable drainage shows the potential for a new relationship with water in cities, as the foundation for a new environmental ethic. Sustainable drainage infrastructure provides the basis for enhanced habitat, localised water management, urban greening and reduced resource intensity. Moving from domination and control to accommodation and partnership brings benefits to human settlements that are impossible using conventional systems. Sustainable drainage shows how starting from natural hydrological and ecological principles as an ethical stance leads to economic, social and health benefits that might otherwise never have been envisaged.

Sustainable drainage maintains discipline over water while acknowledging its role as a positive agent in shaping urban form and experience. Jones and Macdonald (2007) analyse sustainable drainage planning in Glasgow to show the emergence of a more modern form of discipline out of the failure of repressive discipline associated with conventional infrastructure. Drainage practices from the nineteenth century echo pre-modern repression of subjects as a form of disciplining and of controlling behaviour. SuDS in Glasgow offers a more negotiated, modern form of discipline, allowing space for water within the structures and function of the city. The adaptation of drainage towards sustainability required pragmatism in recognising the limits of SuDS and the continued role of grey infrastructure measures, but within an overall shift in approach. Water is no longer seen as a threat to be hidden and removed as quickly as possible, but an active constituent in shaping urban space. This shift towards allowing water into the city in particular spaces reduces the risk of catastrophic failure when water overwhelms the conventional repressive measures. SuDS in Glasgow is far from a revolutionary overthrow of previous repressive disciplining of water, but it is a pragmatic realignment of the relationship between the city and its hydrology towards accommodation.

The repressive disciplining of water as described by Jones and Macdonald (2007) reflects wider patterns of control of nature in industrial, capitalist society. In allowing water, wildlife, trees and wetlands back into the city as functional elements of drainage networks, sustainable drainage approaches represent a shift towards a partnership with nature. Rather than nature being conceived as a resource, a sink for pollutants and a threat to be eliminated or controlled, nature is encouraged in cities. This is far from the 'wild nature' of the Deep Ecologists, but it demonstrates a change in social relationships and cultures away from domination and subordination, as promoted by social ecologists and ecological feminists. Water, plants and soil are active constituents in cities. They are more complex and uncertain than conventional pipes and pumps, but in return offer a host of wider benefits, namely a city that is more habitable for people and other critters.

Drainage for sustainable cities

Sustainable drainage integrates green infrastructure and drainage infrastructure in cities, with potential benefits to surface water management, biodiversity, reduced urban heat island effects, and improved social amenity. Individual technologies and techniques provide multiple benefits. When integrated into high-quality urban design and planning these features can also improve the social and cultural value of the urban environment, including the use of ponds for recreation as well as surface water detention, and the integration of ephemeral streams and swales into public spaces and housing developments.

Sustainable drainage acknowledges the presence of water in the urban environment as a potential benefit rather than a threat. It works to protect and restore local watercourses and wetlands, and to construct them where needed to mimic the pre-development hydrological regime. The need to protect development from the risk of surface water flooding is maintained, and water is understood as a part of the urban environment to be managed for the benefit of people and nature.

References

Ahiablame, L.M., Engel, B.A. and Chaubey, I. 2012. Effectiveness of Low Impact Development Practices: Literature Review and Suggestions for Future Research. *Water Air & Soil Pollution* 223, 4253–4273. doi:10.1007/s11270-012-1189-2

Bean, E.Z., Hunt, W.F. and Bidelspach, D.A. 2007. Field Survey of Permeable Pavement Surface Infiltration Rates. *Journal of Irrigation and Drainage Engineering* 133, 249–255. doi:10.1061/(ASCE)0733–9437(2007)133:3(249)

Bertrand-Krajewski, J-L., Chebbo, G. and Saget, A. 1998. Distribution of Pollutant Mass vs Volume in Stormwater Discharges and the First Flush Phenomenon. *Water Research* 32, 2341–2356. doi:10.1016/S0043–1354(97)00420-X

Bos, J.J. and Brown, R.R. 2012. Governance Experimentation and Factors of Success in Socio-Technical Transitions in the Urban Water Sector. *Technological Forecasting and Social Change* 79, 1340–1353. doi:10.1016/j.techfore.2012.04.006

Braat, L.C. and de Groot, R. 2012. The Ecosystem Services Agenda: Bridging the Worlds of Natural Science and Economics, Conservation and Development, and Public and Private Policy. *Ecosystem Services* 1, 4–15. doi:10.1016/j.ecoser.2012.07.011

Brinkmann, W.L.F. 1985. Urban Stormwater Pollutants: Sources and Loadings. *GeoJournal* 11, 277–283. doi:10.1007/BF00186341

Brown, R., Farrelly, M. and Keath, N. 2009. Practitioner Perceptions of Social and Institutional Barriers to Advancing a Diverse Water Source Approach in Australia. *International Journal of Water Resources Development* 25, 15–28. doi:10.1080/07900620802586090

Brown, R.R. and Farrelly, M.A. 2009. Delivering Sustainable Urban Water Management: A Review of the Hurdles We Face. *Water Science and Technology* 59, 839–846. doi:10.2166/wst.2009.028

Brown, R.R., Farrelly, M.A. and Loorbach, D.A. 2013. Actors Working the Institutions in Sustainability Transitions: The Case of Melbourne's Stormwater Management. *Global Environmental Change* 23, 701–718. doi:10.1016/j.gloenvcha.2013.02.013

Brown, R.R., Keath, N. and Wong, T.H.F. 2009. Urban Water Management in Cities: Historical, Current and Future Regimes. *Water Science & Technology* 59, 847. doi:10.2166/wst.2009.029

Brown, R.R., Sharp, L. and Ashley, R.M. 2006. Implementation Impediments to Institutionalising the Practice of Sustainable Urban Water Management. *Water Science & Technology* 54, 415. doi:10.2166/wst.2006.585

Butler, D. and Davies, J. 2010. *Urban Drainage*, Third Edition. CRC Press, London.

Butler, D. and Parkinson, J. 1997. Towards Sustainable Urban Drainage. *Water Science and Technology, Sustainable Sanitation* 35, 53–63. doi:10.1016/S0273–1223(97)00184–00184

Campisano, A., Butler, D., Ward, S., Burns, M.J., Friedler, E., DeBusk, K., Fisher-Jeffes, L.N., Ghisi, E., Rahman, A., Furumai, H. and Han, M. 2017. Urban Rainwater

Harvesting Systems: Research, Implementation and Future Perspectives. *Water Research* 115, 195–209. doi:10.1016/j.watres.2017.02.056

Cettner, A., Ashley, R., Hedström, A. and Viklander, M. 2014. Sustainable Development and Urban Stormwater Practice. *Urban Water Journal* 11, 185–197. doi:10.1080/1573062X.2013.768683

Chen, S., Liu, H., You, Y., Mullens, E., Hu, J., Yuan, Y., Huang, M., He, L., Luo, Y., Zeng, X., Tang, G. and Hong, Y. 2014. Evaluation of High-Resolution Precipitation Estimates from Satellites During July 2012 Beijing Flood Event Using Dense Rain Gauge Observations. *PLOS ONE* 9, e89681. doi:10.1371/journal.pone.0089681

Costanza, R., d'Arge, R., Groot, R. de, Farber, S., Grasso, M., Hannon, B., Limburg, K., Naeem, S., O'Neill, R.V., Paruelo, J., Raskin, R.G., Sutton, P. and Belt, M. van den, 1997. The Value of the World's Ecosystem Services and Natural Capital. *Nature* 387, 253–260. doi:10.1038/387253a0

Cousins, J.J. 2017. Of Floods and Droughts: The Uneven Politics of Stormwater in Los Angeles. *Political Geography* 60, 34–46. doi:10.1016/j.polgeo.2017.04.002

Coutts, A.M., Tapper, N.J., Beringer, J., Loughnan, M. and Demuzere, M. 2013. Watering Our Cities: The Capacity for Water Sensitive Urban Design to Support Urban Cooling and Improve Human Thermal Comfort in the Australian Context. *Progress in Physical Geography* 37, 2–28. doi:10.1177/0309133312461032

Czemiel Berndtsson, J. 2010. Green Roof Performance Towards Management of Runoff Water Quantity and Quality: A Review. *Ecological Engineering* 36, 351–360. doi:10.1016/j.ecoleng.2009.12.014

Davis, A.P., Hunt, W.F., Traver, R.G. and Clar, M. 2009. Bioretention Technology: Overview of Current Practice and Future Needs. *Journal of Environmental Engineering* 135, 109–117. doi:10.1061/(ASCE)0733–9372(2009)135:3(109)

Davis, A.P., Shokouhian, M., Sharma, H. and Minami, C. 2006. Water Quality Improvement through Bioretention Media: Nitrogen and Phosphorus Removal. *Water Environment Research* 78, 284–293. doi:10.2175/106143005X94376

Davis, A.P., Shokouhian, M., Sharma, H., Minami, C. and Winogradoff, D. 2003. Water Quality Improvement through Bioretention: Lead, Copper, and Zinc Removal. *Water Environment Research* 75, 73–82. doi:10.2175/106143003X140854

DEFRA. 2015. *Sustainable Drainage Systems: Non-Statutory Technical Standards for Sustainable Drainage Systems.* Department for Environment, Food and Rural Affairs, London.

Dover, J.W. 2015. *Green Infrastructure: Incorporating Plants and Enhancing Biodiversity in Buildings and Urban Environments.* Routledge, London and New York.

Ellis, J.B. 2013. Sustainable Surface Water Management and Green Infrastructure in UK Urban Catchment Planning. *Journal of Environmental Planning and Management* 56, 24–41. doi:10.1080/09640568.2011.648752

Farrelly, M. and Brown, R. 2011. Rethinking Urban Water Management: Experimentation as a Way Forward? *Global Environmental Change, Special Issue on the Politics and Policy of Carbon Capture and Storage* 21, 721–732. doi:10.1016/j.gloenvcha.2011.01.007

Finewood, M.H. 2016. Green Infrastructure, Grey Epistemologies, and the Urban Political Ecology of Pittsburgh's Water Governance. *Antipode* 48, 1000–1021. doi:10.1111/anti.12238

Fitzgerald, J. and Laufer, J. 2017. Governing Green Stormwater Infrastructure: The Philadelphia Experience. *Local Environment* 22, 256–268. doi:10.1080/13549839.2016.1191063

Hedgcock, D. and Mouritz, M. 1993. Water Sensitive Residential Design. *Australian Planner* 31, 114–118. doi:10.1080/07293682.1993.9657618

Jiang, Y., Zevenbergen, C. and Fu, D. 2017. Understanding the Challenges for the Governance of China's "Sponge Cities" Initiative to Sustainably Manage Urban Stormwater and Flooding. *Natural Hazards* 89(1), 521–529. doi:10.1007/s11069-017-2977-1

Jones, P. and Macdonald, N. 2007. Making Space for Unruly Water: Sustainable Drainage Systems and the Disciplining of Surface Runoff. *Geoforum, Post Communist Transformation* 38, 534–544. doi:10.1016/j.geoforum.2006.10.005

Kambites, C. and Owen, S. 2006. Renewed Prospects for Green Infrastructure Planning in the UK. *Planning Practice & Research* 21, 483–496. doi:10.1080/02697450601173413

Kang, C.D. and Cervero, R. 2009. From Elevated Freeway to Urban Greenway: Land Value Impacts of the CGC Project in Seoul, Korea. *Urban Studies* 46, 2771–2794. doi:10.1177/0042098009345166

Karvonen, A. 2010. Metronatural™: Inventing and Reworking Urban Nature in Seattle. *Progress in Planning* 74, 153–202. doi:10.1016/j.progress.2010.07.001

Karvonen, A. 2011. *Politics of Urban Runoff: Nature, Technology, and the Sustainable City*. MIT Press, Cambridge, MA.

Kayhanian, M. and Stenstrom, M. 2005. Mass Loading of First Flush Pollutants with Treatment Strategy Simulations. *Transportation Research Record: Journal of the Transportation Research Board* 1904, 133–134. doi:10.3141/1904-1914

Loperfido, J.V., Noe, G.B., Jarnagin, S.T. and Hogan, D.M. 2014. Effects of Distributed and Centralized Stormwater Best Management Practices and Land Cover on Urban Stream Hydrology at the Catchment Scale. *Journal of Hydrology, Water Governance Across Competing Scales: Coupling Land and Water Management Incorporating Water Resources in Integrated Urban and Regional Planning* 519, Part C, 2584–2595. doi:10.1016/j.jhydrol.2014.07.007

Melville-Shreeve, P., Ward, S. and Butler, D. 2016. Rainwater Harvesting Typologies for UK Houses: A Multi Criteria Analysis of System Configurations. *Water* 8, 129. doi:10.3390/w8040129

Mitchell, V.G. 2006. Applying Integrated Urban Water Management Concepts: A Review of Australian Experience. *Environmental Management* 37, 589–605. doi:10.1007/s00267-004-0252-1

Mugume, S.N., Melville-Shreeve, P., Gomez, D. and Butler, D. 2016. Multifunctional Urban Flood Resilience Enhancement Strategies. *Proceedings of the Institution of Civil Engineers – Water Management* 170, 115–127. doi:10.1680/jwama.15.00078

Niemczynowicz, J. 1999. Urban Hydrology and Water Management – Present and Future Challenges. *Urban Water* 1, 1–14. doi:10.1016/S1462-0758(99)00009-00006

Parikh, H. and Parikh, P. 2009. Slum Networking – A Paradigm Shift to Transcend Poverty with Water, Environmental Sanitation and Hidden Resources, in: *Water and Urban Development Paradigms: Towards an Integration of Engineering, Design and Management Approaches*. CRC Press, London, pp. 357–370.

Parkinson, J., Tayler, K. and Mark, O. 2007. Planning and Design of Urban Drainage Systems in Informal Settlements in Developing Countries. *Urban Water Journal* 4, 137–149. doi:10.1080/15730620701464224

PWD. 2009. *A Triple Bottom Line Assessment of Traditional and Green Infrastructure Options for Controlling CSO Events in Philadelphia's Watersheds*. Philadelphia Water Department, Philadelphia, PA.

PWD. 2011. *Amended Green City Clean Waters: The City of Philadelphia's Program for Combined Sewer Overflow Control*. Philadelphia Water Department, Philadelphia, PA.

PWD. 2016. *5 Down, 20 to Go: Celebrating 5 Years of Cleaner Water and Greener Neighborhoods* [WWW Document]. Philadelphia Water Department. http://phillywatersheds.org/5Down.

Ranganathan, M. 2015. Storm Drains as Assemblages: The Political Ecology of Flood Risk in Post-Colonial Bangalore. *Antipode* 47, 1300–1320. doi:10.1111/anti.12149

Rijke, J., Farrelly, M., Brown, R. and Zevenbergen, C. 2013. Configuring Transformative Governance to Enhance Resilient Urban Water Systems. *Environmental Science & Policy* 25, 62–72. doi:10.1016/j.envsci.2012.09.012

Sundaresan, J., Allen, A. and Johnson, C. 2017. Reading Urban Futures Through Their Blue Infrastructure: Wetland Networks in Bangalore and Madurai, India, in: *Urban Water Trajectories, Future City*. Springer, Cham, pp. 35–50. doi:10.1007/978-3-319-42686-0_3

Teemusk, A. and Mander, Ü. 2007. Rainwater Runoff Quantity and Quality Performance from a Greenroof: The Effects of Short-Term Events. *Ecological Engineering* 30, 271–277. doi:10.1016/j.ecoleng.2007.01.009

Valderrama, A., Levine, L., Bloomgarden, E., Bayon, R., Wachowitz, K. and Kaiser, C. 2013. *Creating Clean Water Cash Flows – Developing Private Markets for Green Stormwater Infrastructure in Philadelphia*. Natural Resources Defense Council, New York City.

Voskamp, I.M. and Van de Ven, F.H.M. 2015. Planning Support System for Climate Adaptation: Composing Effective Sets of Blue-Green Measures to Reduce Urban Vulnerability to Extreme Weather Events. *Building and Environment, Special Issue: Climate Adaptation in Cities* 83, 159–167. doi:10.1016/j.buildenv.2014.07.018

Woods Ballard, B., Wilson, S., Udale-Clarke, H., Illman, S., Scott, T., Ashley, R. and Kellagher, R. 2015. *The SuDS Manual (No. CIRIA C753)*. CIRIA, London.

8 Reuse

Introduction

Reuse and recycling are familiar strategies for reducing the resource intensity of cities. Water is the original recyclable material. The global water cycle has been recycling water around the planet for billions of years. Water reuse is unsurprisingly a key component of discussions about urban water sustainability.

Modern cities treat water as a disposable product. It enters through the drinking water network, is used for a matter of seconds, and then disappears through the wastewater network. Water that falls on the city as rainfall is drained away as quickly as possible, and is treated as a hazard, not a resource. This linear pattern of water infrastructure has been criticised by proponents of sustainability, particularly those who look upon cities from an industrial ecological view of urban metabolism (Castán Broto et al., 2012; Hermanowicz and Asano, 1999). The industrial ecology analysis of urban metabolism takes account of the material inputs and outputs of cities and looks for opportunities to increase material productivity by reusing and recycling materials. For water, this means moving from linear to circular patterns of flow, whereby water is used more than once within the city.

The continuous supply of clean water throughout the city has been essential to delivering good public health. However, as water supplies are limited and the energy required to treat water and wastewater increases, the logic of using clean drinking water for flushing toilets, watering gardens and other low-contact or non-critical functions is questionable (Hermanowicz and Asano, 1999). The concept of 'fit-for-purpose' water systems acknowledges that many water demands, such as for toilet flushing and gardening, can be met using water that does not meet drinking water standards (Barton and Argue, 2009; Hurlimann and McKay, 2006; Mitchell, 2006; Mitchell et al., 2002). This allows for more diverse sources and providers of water to be integrated into urban water systems, including household greywater reuse and rainwater harvesting, and neighbourhood-scale dual-reticulation schemes providing both potable and recycled, non-potable water to households and businesses (Brown et al., 2009). Municipal wastewater provides a potential resource under conditions of water scarcity. Wastewater is used for either potable or non-potable purposes, with implications for the technology and energy used in treatment, risk management and public acceptability.

Water reuse takes many different forms, uses many different technologies and operates at different scales (Lazarova et al., 2013; McClelland et al., 2012; Tjandraatmadja et al., 2005). The different forms of water reuse open up the possibility for infrastructure systems and technologies that do not need the high standards of treatment required for drinking water, particularly if the source of water for such uses is relatively clean, compared to the reuse of highly contaminated municipal wastewater (Jefferson et al., 2004; Toze, 2006). A centralised system may be required to continuously and reliably deliver high-quality, safe water at low risk for the public to drink, or for reuse of wastewater. However, for less risky water uses from less contaminated sources, lower levels of control and treatment are required, allowing for more distributed approaches to the supply and management of water.

For many people reuse and recycling are a central tenet of sustainability, yet water recycling is not inherently sustainable. In some forms reuse can increase, rather than decrease, environmental impacts relative to conventional water systems. The sustainability of different configurations of water reuse is judged in terms of reduction in demand for water resources, local environmental impact, public health security, social acceptability, cost and energy demand. Water recycling can be costly, risky, energy intensive and controversial due to public concerns about health and environmental risks (Makropoulos and Butler, 2010; Memon et al., 2005). Water reuse presents complex choices and implementation challenges to urban water managers.

This chapter addresses the different options for water recycling in cities. It presents the different forms of potable and non-potable reuse and implications for energy consumption, social acceptability and technology choice. The environmental impacts of different forms of reuse are reviewed before reuse is analysed using the five frameworks of sustainability.

Forms of reuse

Technologies and systems for reusing wastewater and harvesting rainwater have been part of daily practice and urban infrastructure since ancient times. In modern cities, water reuse has been proposed as a strategy for overcoming water shortages and improving sustainability since the 1960s. The different forms of water reuse in modern cities can categorised using binary distinctions: potable – non-potable; direct – indirect; planned – unplanned; and centralised – decentralised (Wilcox et al., 2016). Potable reuse uses treated wastewater as a source of supply for the conventional drinking water system (Bell and Aitken, 2008). Non-potable reuse involves a separate system for non-potable water use, and ranges from householders using dishwater to water houseplants to elaborate 'purple-pipe' dual-reticulation schemes that deliver treated municipal wastewater to homes as a second water supply service, in addition to the standard potable water infrastructure (Li et al., 2010; Okun, 1997). Potable reuse can be direct, where treated wastewater is connected to the drinking water treatment system, or indirect, where the water to be reused is discharged into the environment or raw water storage such as a reservoir, local aquifer or shortly upstream of a river intake (Guo and Englehardt, 2015; Leverenz et al., 2011). Potable reuse can be planned, where treated wastewater is deliberately reintroduced into the drinking water supply system, or unplanned, most typically in urbanised catchments where wastewater discharge into a river from one city contributes to the supply of a downstream city. Potable reuse is most commonly centralised, using municipal wastewater and potable distribution networks, but it can be decentralised, providing drinking water at building or neighbourhood scale. Non-potable reuse is more typically planned and direct, but can be centralised, through dual-reticulation schemes, or decentralised, at the household, building or neighbourhood scale.

Water reuse can also be characterised by the source of the water – municipal wastewater, industrial wastewater, greywater, rainwater or stormwater. The level of treatment required between source and use is determined both by the quality of the water being reused and the quality required for the end use (Toze, 2006). Assessment of risk to public health and the environment is a core element of decisions about the required end-use quality and the treatment technology. In modern cities risk management and public perception concerns tend to require high levels of treatment to avoid public health risks and improve public acceptability (Mankad, 2012; Toze, 2006).

The dirtier the source of wastewater and the higher the quality required for the end use, the higher the level of treatment required and the greater the associated energy consumption (Bixio et al., 2008). Municipal wastewater treatment is typically the most energy-intensive element of conventional urban water infrastructure (McCarty et al., 2011; Venkatesh and Brattebø, 2011). Treating municipal wastewater to a quality sufficient for potable or non-potable reuse is the most energy-intensive reuse option, but much of this energy would be expended in any case to treat the wastewater for safe release into the environment (Bixio et al., 2008; Joss et al., 2008). If water is to be treated to a high standard for release into the environment, the additional energy required to treat and distribute the water for reuse may compare favourably to alternative options for new water supply such as long-distance transfers and desalination (Lim et al., 2010; Tangsubkul et al., 2005).

The scale of reuse also influences energy requirements, with decentralised systems theoretically requiring less energy for distribution and lower pressure losses due in distribution networks, to be traded off against the efficiencies of scale achievable with larger treatment works, pumps and distribution networks (Gikas and Tchobanoglous, 2009; Livingston et al., 2004). The relative inefficiency and overdesign of small-scale treatment and pumping for distribution can mean that decentralised reuse is more energy intensive than centralised drinking water supply (Crettaz et al., 1999).

Potable reuse

In cities facing long-term water scarcity, potable reuse is often considered alongside desalination as a potential new source of water. Potable reuse has been practiced since the 1970s in places including Windhoek, Namibia, and Orange County in the United States (du Pisani, 2006; Mills and Watson, 1994). Early potable reuse schemes were based on conventional wastewater treatment with additional chemical removal of nutrients and contaminants. More recently potable reuse has benefited from developments in membrane technologies, and most potable reuse schemes proposed or implemented since the 1980s involve reverse osmosis (RO) treatment of treated wastewater prior to either direct or indirect reintroduction into the drinking water treatment system. RO is the same technology used for desalination, discussed in detail in Chapter 9. Potable reuse therefore has high energy requirements relative to conventional treatment, but the low salinity of treated municipal wastewater compared to brackish water or seawater means that it is less energy intensive than desalination (Cooley and Wilkinson, 2012; Stokes and Horvath, 2006).

Potable recycling is less energy intensive but more socially controversial than desalination because of public concerns about risks associated with recycling sewage (Bridgeman, 2004; Dolnicar et al., 2011; Hartley, 2006; Marks, 2006). Indirect reuse has been preferred to direct reuse as it is considered to be less controversial because of reduced risks of micro-contaminants recirculating in the drinking water system. Direct potable reuse, where treated wastewater is returned to the drinking water treatment and distribution without an environmental buffer, is used in Windhoek and has recently been implemented in Big Spring and Wichita Falls in Texas in the US and in Beaufort West, South Africa (du Pisani, 2006; Lahnsteiner et al., 2017).

Planned potable reuse schemes are currently operating in the US, Belgium, Germany, Spain, Namibia, South Africa, the UK and Singapore (Angelakis and Durham, 2008; Lazarova et al., 2013). The expansion of potable reuse has been constrained by concerns about public acceptability. Concerns about potential health risks from reusing wastewater formed the basis of significant controversies about potable reuse in Australia and the US in the 1990s and 2000s (Aitken et al., 2014; Hartley, 2006; Marks et al., 2008). Potable recycling has a contentious history in terms of public acceptability, with notable cases including the city of Toowoomba in Australia where a proposition for water recycling was rejected in a referendum in 2006 (Box 8.1), and San Diego in 2000 where a newly commissioned plant was shut down while public concerns about the environmental justice of poorer communities being supplied with recycled water were addressed (Bridgeman, 2004; Hurlimann and Dolnicar, 2010; Price et al., 2012).

Recent experiences of extended drought in the southern US and South Africa have shown that the public are willing to accept potable reuse in extreme water shortages. Wichita Falls and Deep Spring, Texas, have

Box 8.1   Potable reuse referendum in Toowoomba, Australia

Public acceptability is widely recognised as a key consideration in pursuing potable reuse, but existing decision-making structures have struggled to account for public deliberation (Colebatch, 2006; Hartley, 2006). The City of Toowoomba in Australia held a referendum on indirect potable water reuse as a solution to water shortages during an extended drought in 2006. The referendum was a condition of receiving funding for the project from the Australian federal government and was opposed by the local government, who wanted to implement the scheme without delay. After a short and divisive campaign, 62% of citizens voted against the proposal for IPR. As the drought continued, the state government of Queensland announced that Toowoomba was to be supplied from a regional network, which included recycled water from another catchment. Two years after the referendum residents' attitudes to drinking recycled water had softened, partly as a result of necessity. The referendum forced debate into adversarial argument for or against a particular technology, leaving little room for addressing public concerns or evaluating alternatives within the design and decision-making process. Whilst the public expressed legitimate concerns, these were not able to be addressed adequately in the design and implementation of the proposed system due to the adversarial nature of the referendum and the disjointed governance approach across local, state and federal jurisdictions (Hurlimann and Dolnicar, 2010).

implemented direct potable reuse as a means of maintaining supply to local homes and businesses. In these inland locations desalination is not viable, and communities have shown acceptance of direct reuse in contrast to conventional expectations that indirect reuse is required for community acceptance. Perth in Western Australia has implemented indirect potable reuse by using treated wastewater to augment local groundwater resources. In this case potable reuse is implemented alongside desalination as part of a strategy of 'climate-proofing' water supply. In London potable reuse is proposed as the primary source of water to address future supply-demand deficits, while desalination remains a resilience measure to be used only in times of severe drought.

Non-potable reuse

Non-potable reuse varies greatly in the sophistication and energy intensity of treatment. Treatment technologies are of variable complexity, depending on the source of water and how it is stored and used. Non-potable supply can be managed by laypeople in their own homes and gardens, by building owners and landlords, by third-party service and technology providers, and by water utilities.

Household-scale non-potable reuse can be as simple as manually bucketing bathwater into the garden or toilet, but is typically discussed in terms of greywater. Domestic greywater reuse systems can start from simple redesign of plumbing fixtures, such as installing a hand basin above the cistern of a toilet to reuse handwashing water for flushing. More elaborate household-scale greywater reuse takes water from the shower and washing machine, through a simple filter and settling tank, with some systems dosing with disinfectant, and final recirculation to the toilet or garden. The energy requirements of domestic greywater systems can depend significantly on the layout of the system, and arrangements that utilise gravity instead of mechanical pumping require lower energy input. Domestic rainwater reuse operates similarly to greywater reuse, with the water collected from roof downpipes filtered, stored and pumped to non-potable end uses.

Membrane bioreactors (MBR) are commonly used in decentralised treatment of wastewater, and for non-potable reuse of municipal wastewater (Verrecht et al., 2012). Membrane bioreactors allow for biological degradation and filtering of wastewater in a single reactor unit, significantly reducing the physical footprint required for wastewater treatment (Melin et al., 2006; Wisniewski, 2007). The incorporation of membrane filtration in the wastewater treatment process also allows for removal of viruses and dissolved contaminants that are not reliably removed by conventional wastewater treatment. Membrane bioreactors are more energy intensive than conventional wastewater treatment; however, energy use overall in reuse systems compared to the whole system requirements for energy from conventional, linear systems can be comparable or beneficial (Tangsubkul et al., 2005; Verrecht et al., 2012). Non-potable water reuse may also use conventional, non-membrane tertiary treatment processes, such as filtration and flocculation, and may also include disinfection. The final water quality may be similar to potable water standards to minimise public health risks from cross-connection or misuse of the recycled water.

Non-potable recycling of wastewater uses water that would otherwise be discharged to the environment, which has been treated to a standard that is safe for human contact but not human consumption (Gikas and Tchobanoglous, 2009). Irrigation of municipal playing fields, gardens and golf courses with wastewater discharge has been practiced in Australia and the US for several decades, and Israeli agriculture is highly dependent on treated wastewater (Asano and Levine, 1996; Friedler, 2001).

Dual-reticulation reuse systems provide treated wastewater for non-potable and drinking water for potable uses in separate distribution networks. The first dual distribution system in the US was built in the 1920s to supply Grand Canyon Village in Arizona (Okun, 1997). Early discussions of integrated urban water management in the 1960s by engineers such as Daniel Okun included propositions for and analysis of dual-reticulated reuse systems (Okun, 1973; Okun et al., 1969). More recently in Europe and Australia housing and public space developments have been built with dual plumbing systems to supply drinking water and recycled water for non-potable use (Guo and Englehardt, 2015; Lazarova et al., 2003; Willis et al., 2011). Other sources of water that may be treated for non-potable reuse include 'sewer mining' and 'stormwater mining', where water is abstracted from drainage networks (Hatt et al., 2006; Mitchell et al., 2002).

Non-potable dual-reticulation developments were implemented in the 2000s in Australia in response to an extended drought. The Pimpama Coomera development on the Gold Coast included a dual-reticulation system for non-potable reuse, owned by the local water utility (Radcliffe, 2015). The scheme operated from 2008 until 2016 when it was decommissioned due to high operating costs. In London a dual-reticulation system was built to supply water for landscape irrigation and toilet flushing for the 2012 Olympic Games (Verrecht et al., 2012). Thames Water has since operated the system as a demonstration site, to improve understanding of the MBR technology and associated systems, but the commercial viability of the plant is highly vulnerable as it operates at much higher costs than conventional supply and wastewater treatment. Dual-reticulation schemes therefore risk becoming stranded assets if decisions are made to subsidise otherwise uneconomic forms of water supply during drought.

Greywater reuse and rainwater harvesting have been encouraged and mandated in planning policy and building rating schemes in some jurisdictions. Green building certification systems such as the US-based LEED, the UK-based BREEAM and the New South Wales BASIX schemes all include greywater and rainwater harvesting as contributing towards higher eco-building ratings. However, evaluation of rainwater harvesting systems has shown overestimation of the water savings achieved, indicating that these systems need to be better managed and integrated into building operation, maintenance and water management programmes (Moglia et al., 2016). Integrating user and household needs into the design and operation of decentralised water reuse options is important to achieve the full water supply benefit.

The water-saving potential of domestic non-potable reuse systems may be undermined if the systems are topped up by mains water to ensure consistent supply as is required by the relevant British Standards (BSI, 2010). In the first dual-reticulated housing development in Australia at Rouse Hill in Sydney, overall consumption of water increased when the scheme was first implemented, highlighting the importance of integrating water conservation and pricing strategies within water recycling schemes (Law, 1996; Livingston et al., 2004).

A risk-based approach to regulation of household and building-scale recycling and harvesting of water is emerging, to recognise the variable complexity of technology and management systems. For instance, The *Code of Practice for the Reuse of Greywater in Western Australia 2010* recognises different levels of technology and complexity in reuse systems, and applies variable levels of regulation of technology and use (Department of Health, 2010). Simple buckets and pipes used to redistribute laundry water for gardening can be used for immediate surface irrigation of gardens without storage, and do not require external monitoring or approval. Systems that store water for more than 24 hours must be provided by an approved supplier and can only be used for subsurface irrigation of gardens or for plumbed non-potable use. The requirement for approval of technologies has been criticised as increasing the cost of domestic reuse systems, restricting DIY innovation and encouraging complex treatment and pumping systems which require higher energy use. Requirements for subsurface irrigation have also been criticised as increasing costs unnecessarily in local environments with freely draining soils which are unlikely to pose any additional health risk from surface ponding of greywater.

Environmental impacts

Whilst saving water, rainwater harvesting and greywater reuse can be more energy and greenhouse gas intensive than conventional supply and reuse of municipal wastewater (Stokes and Horvath, 2006). Rainwater harvesting has been shown to be more energy intensive than mains water for domestic reuse, including for toilet flushing (Crettaz et al., 1999; Parkes et al., 2010). Energy use is highly dependent on the configuration of the system and the level of treatment desired. For rainwater harvesting, energy use can be reduced by minimising pumping requirements, repositioning storage tanks and reducing requirements for recirculation of water.

Compared to potable reuse systems, dual-reticulation networks require additional embodied energy and materials to build the second distribution network and additional pumps to supply the water to end users. Where these systems are providing water that is close to potable standard to avoid public health risk, the principle of 'fit-for-purpose' may be undermined and the energy, material and economic costs of building and running a separate non-potable distribution network may outweigh potential benefits of non-potable over potable reuse or other potable water supply options (Lundie et al., 2004; Tangsubkul et al., 2005).

Reusing greywater reduces the overall volume and increases the contaminant concentration of urban wastewater. As well as reducing pressure on water resources, water reuse also reduces wastewater discharge to the environment. The Rouse Hill dual-reticulation scheme in Sydney was partially motivated by limits on discharge of wastewater to local rivers (Cooper, 2003; Law, 1996). In recirculating water in the city, water abstractions and wastewater discharge to the environment are both reduced. At a large scale, reducing the volume of wastewater could require changes in the operation of sewers and wastewater treatment works, and may contribute to increased sewer blockages in poorly maintained domestic drains.

Framing reuse

Water reuse demonstrates that even a technology that seems inherently sustainable involves trade-offs and counter-intuitive impacts. Concerns about public acceptability also highlight the importance of social impacts and public engagement in urban water sustainability. The sustainability of reuse schemes can be quantitatively assessed using metrics of energy use, water saved and water quality impacts, but its wider role in urban sustainability differs according to different theoretical and policy frameworks.

Sustainable development

Water reuse is consistent with the familiar focus on recycling in sustainable cities discourse. It is an important element of strategies for integrated water management to meet Sustainable Development Goal 7, including reuse for agriculture as well as a supply option for cities. Water reuse and rainwater harvesting may be included in appropriate technology approaches to reducing urban water poverty, providing low-capital, decentralised alternatives to centralised provision of water and wastewater services (Campisano et al., 2017; Handia et al., 2003; Mandal et al., 2011; Srinivasan et al., 2010). Informal reuse can be a strategy for households to deal with water poverty and to improve livelihoods through irrigation, but it requires appropriate policies and technologies to manage risks to health and the environment (WHO, 2006).

Urban water reuse provides opportunities to address the water-energy-food nexus within local urban environments. Informal and formal reuse of water for irrigation in urban and peri-urban agriculture supports urban livelihoods and productivity in cities in the Global South (Mahesh et al., 2015; Makoni et al., 2016; Miller-Robbie et al., 2017). Agricultural reuse benefits from nutrients in wastewater, increasing productivity and reducing demand for fertiliser. However, agricultural workers exposed to untreated wastewater face significant health risks (Ensink et al., 2002; Huertas et al., 2008; Qadir et al., 2010). Reusing water for agriculture requires treatment to minimise risks to workers, the public, the environment and consumers of agricultural products.

Building new infrastructure as part of urban development and regeneration programmes provides opportunities to implement water systems that include more circular flows of water and utilised local water sources, including greywater and treatment municipal water. Cost and energy consumption are key considerations in the long-term feasibility and sustainability of such schemes, as evidenced by the decommissioning of the Pimpama Coomera dual-reticulation scheme in Australia (Radcliffe, 2015).

Water reuse is promoted as a resilience measure through diversifying water sources and providing an additional, climate-independent source of water (Ferguson et al., 2013; Wong and Brown, 2009). Water reuse expanded rapidly in Australia during the drought at the beginning of the twenty-first century to provide new sources of water, and drought resilience is a motivation for individuals as well as utilities and cities to invest in water reuse systems (Radcliffe, 2015). Resilience may also be improved through connections within a system, across centralised and decentralised networks. Hybrid approaches to water sustainability and resilience are emerging which address the role of different scales of water infrastructure, including different scales and approaches to water reuse (Hwang et al., 2014; Makropoulos and Butler, 2010; Sapkota et al., 2014).

Ecological modernisation

Ecological modernisation approaches to water reuse draw attention to the possibility for technological innovation to deliver new sources of water at different scales. Reuse technologies provide options to substitute wastewater of varying qualities for conventional water resources, without disrupting modern lifestyles and urbanisation processes. New technologies represent business opportunities, but the uptake of reuse, particularly decentralised systems, is constrained by its high cost relative to existing infrastructure provision.

Potable reuse presents a technical solution to water scarcity. Water becomes an industrial product that can be endlessly recycled, as membranes produce freshwater faster and more reliably than natural processes of hydrology. Ecological modernist approaches to potable reuse focus on improving the safety, efficiency and reliability of the technology, but have been challenged by concerns about public acceptability. Discussions of 'public acceptability' from an ecological modernisation standpoint imply that the technology itself is sound, and that the public are irrational in 'rejecting' a modern water source. Public concerns are thought to arise from irrationality, such as disgust, traditional religious beliefs and lack of technical knowledge. Efforts to engage the public in decisions about potable reuse have evolved to account for legitimate concerns and to position reuse within a range of options for water supply. However, campaigns based on educating the public about the safety and benefits of reuse technologies in order to convince them to accept this new form of water supply reinforce expert-led decision-making about water infrastructure. Potable reuse is constructed as a modern form of water supply, and public concerns are barriers to progress that need to be overcome in order to deliver sustainable water systems.

Dual-reticulation reuse schemes are a technically innovative form of infrastructure, but their sustainability is undermined by high costs. Water reuse schemes that are implemented for demonstration purposes or during drought have struggled for longer-term viability compared to conventional water supplies. Such schemes have provided important opportunities for research and development, performance evaluation and technical innovation. Whilst they may reduce demand for conventional water resources and fulfil environmentalist impulses to recycle materials and work towards a circular economy, they are not yet economically competitive with conventional supplies without significant subsidy. Further innovation in business models and institutional reform are needed to provide a context for the technical innovations to succeed.

Non-potable reuse at household scale provides an alternative means for water users to meet their needs without depending on the potable mains supply. Whilst this can reduce demand on regional water supplies, it can be used to maintain high-water-using lifestyles. Private individuals who invest in rainwater harvesting or recycling systems may reflect a strong environmental ethic, or they may be less inclined to reduce their overall water use. Alternative supplies can be used to avoid reducing, undermining wider transitions to sustainability.

Sustainable building codes rating schemes have incorporated water reuse within a particular tool for improving resource efficiency in cities. Such schemes are often voluntary and enable developers' flexibility in how they achieve various standards of environmental performance. Water reuse is usually only required to achieve the highest levels of rating. The energy requirements associated with water reuse systems can increase the overall energy use of the building, in conflict with green building code goals of reducing energy use. This has led to calls for greater regional flexibility for the water reuse elements of building assessment, to avoid unnecessarily increasing energy use by installing greywater and rainwater systems in buildings in regions with abundant water supplies, and to promote water reuse in water-scarce regions. Such flexibility is consistent with ecological modernisation policy approaches, to enable technological innovation and market-led solutions to local and global environmental problems.

Socio-technical systems

A socio-technical approach to water reuse emphasises the need to address social, institutional and governance structures and processes alongside technical developments. Controversies over potable reuse demonstrate the need for new decision-making processes for water infrastructure, decentralised reuse requires new regulations and business models, and household-scale reuse systems must be embedded within everyday life and water-using practices to be successfully implemented over the longer term.

Potable reuse represents minimal disruption to conventional infrastructure arrangements, providing a new source of supply to meet growing demand or to replace over-abstracted water resources. The success of potable reuse projects depends largely on public acceptability, which varies depending on the level of public engagement during project development, the point of reintroduction of the recycled water into the drinking water resource (groundwater, surface water or reservoir), and cultural differences between countries and communities. Conventional strategies for managing public controversy about reuse are based on a deficit model of the public understanding of science and technology (Sturgis and Allum, 2004). This holds that public rejection of technology is due to ignorance, and consequently that public education about the science and engineering of the technology and its benefits will increase acceptability. Whilst there is some evidence that greater understanding of water systems can lead to higher levels of trust and acceptance of new technologies, the deficit model fails to meaningfully engage with legitimate public concerns or to involve the public in complex decisions about future water supplies (Colebatch, 2006).

Socio-technical approaches to the development of new technologies and infrastructures for urban water sustainability promote deliberative processes that allow for public concerns and ideas to be accounted for in engineering design and decision-making (Bell and Aitken, 2008). Public controversies about water reuse show that communities are less willing to accept water infrastructure decisions made by experts than in earlier periods of infrastructure development. Deliberative approaches to engaging with the public in decisions about controversial technologies, such as potable water reuse, are promoted from a socio-technical viewpoint as enabling greater democratic accountability and oversight of decisions (Bell and Aitken, 2008; Russell and Lux, 2009).

Socio-technical analysis of non-potable reuse emphasises the need for clear regulation, user engagement and business models to support the dissemination and maintenance of technology. Domestic water reuse that requires intervention by householders or changes to daily practice and routines are less likely to achieve sustainable reductions in water demand than interventions that are embedded in existing daily practice or have been designed and implemented with user needs in mind. Technologies and systems need to be designed, implemented, managed and maintained in specific contexts, requiring social, economic and institutional support to succeed. In Melbourne rainwater harvesting tanks have been widely installed and promoted through building regulations and rebates, but only one in four systems are operational, emphasising the need to design programmes and technologies that account for user needs and expertise (Moglia et al., 2016).

Everyday habits of water reuse, such as reusing dishwater on a kitchen garden or diverting washing-machine water onto a lawn, can provide the basis for design of supportive technologies. Zoe Sofoulis's work in Sydney highlighted the saver-unfriendliness of most domestic water-using appliances and fittings (Sofoulis, 2005). Technologies that are designed for a linear, once-through flow of water are difficult for users to adapt to enable domestic reuse during times of water scarcity. While users may value water saving and reuse, household technologies and plumbing are 'locked-in' to conventional infrastructures of supply and wastewater disposal. This makes water recycling practice difficult, but domestic innovators with high motivation to reuse water develop workarounds that could be the basis for future engineering design to support reuse. Rather than simply providing the technology needed for water reuse and expecting users to develop compliant behaviours, designing technology to support nascent water reuse practices or to fulfil user desires for more sustainable water sources could lead to longer-term transition to sustainable domestic water systems.

According to transitions theory, water reuse remains a niche innovation which has yet to be widespread because the socio-technical regime and landscape for water infrastructure have remained stable (Geels, 2002). Water reuse has been incorporated into the conventional governance and ownership of water infrastructure, particularly potable reuse and dual-reticulated non-potable reuse networks. However, these schemes remain vulnerable as they are more costly than conventional infrastructure operating at a larger scale. While the regime and landscape for urban water infrastructure is dominated by the institutional, economic and regulatory forms that have evolved to support centralised, linear water infrastructure, widespread diffusion of water reuse is unlikely.

Political ecology

Analysing water reuse as a solution to the socio-environmental problem of water security highlights inequalities in the distribution of benefits, costs and risks associated with implementing new infrastructure options. Potable reuse maintains existing infrastructural power relationships but may have different real or perceived impacts on different social groups. Non-potable reuse may also have different social impacts. Wealthier homeowners are more likely to be able to afford to install, maintain and operate decentralised greywater and rainwater systems and take advantage of government subsidies than households on lower incomes or in rental accommodation.

Environmental justice analysis points to the disproportionate impact of environmental problems on poor and marginalised communities. Poorer communities in cities are often more exposed to environmental pollution and more vulnerable to natural disasters. Public opposition to potable reuse in San Diego in the 1990s and 2000s included concerns from African American residents about the potential for disproportionate exposure to risk compared to wealthier, white communities. Environmental justice points out that in potable reuse schemes marginalised people, who on average consume less water than wealthier people, may be exposed to unknown risks associated with implementing a new technology, while more affluent residents continue to receive water from well-established sources using mature treatment technologies. While the risk of harm may technically be very small, the burden of risk falls on the most vulnerable residents of the city, raising concerns about the justice of distribution of risks and benefits from the new technology. Engaging communities in decision-making and addressing concerns about distribution of water and risks is important to ensure fairness in the selection and implementation of new potable water sources.

Decentralised water reuse owned and operated by households and communities represents alternate economic models to the dominant capital-intense infrastructure provision. Redistributing water supply through water reuse technologies also redistributes ownership and governance. Household and community provision of non-potable water could be seen as further withdrawal of the state from providing basic, universal public services. Alternately, it can support and activate local economic development and empowerment. However, decentralised models of ownership of infrastructure should not be valorised as inherently fair and sustainable, as they may be vulnerable to capture by powerful actors, corruption, inefficiency and poor technical capacity (Bakker, 2010).

Radical ecology

Radical ecology seeks to reduce human impact on the natural world through minimising resource use and disruption to local landscapes and ecosystems. Judicious use of resources taken from the environment is central to this position, and so water reuse is essentially consistent with a radical ecology approach to the relationship between the city and the natural world. Radical ecology encourages living within local ecosystems and landscapes, and it has been associated with decentralised approaches to politics and technology, as a means of reducing impacts and of helping people to develop sustainable and sustaining relationships with the natural world.

Radical ecologists would therefore promote low-energy, small-scale water reuse technologies, particularly systems that help users to a better understanding of water as part of nature as well as a human right and resource. The appropriate technology movement and its 'small is beautiful' approach uses local resources to build simple systems that can be operated and maintained at low cost by non-experts. Household or communal-scale greywater reuse and rainwater harvesting are well-established appropriate technologies, emphasising self-reliance or community-based management of technology and resources rather than dependence on large utilities and complex technological infrastructures. Such 'off-grid' systems require energy-efficient design, as access to energy is also constrained if the water system is part of a wider decentralised approach to resource management and infrastructure.

Domestic-scale water reuse and rainwater harvesting can be evidence of a heightened environmental ethic and intention to reduce individual impacts on shared resources and ecosystems. It can also be used to maintain high water-consuming practices and lifestyles during times of resource scarcity. In this case water reuse can be seen to provide an increasingly complex technological buffer between water users and the natural environment, rather than adapting practices and technologies to local conditions. Water reuse as a liberal, individualist technology enabling the personal freedom to continue to exploit resources can undermine efforts to establish a greater shared environmental ethic.

In contrast to decentralised systems, potable reuse and utility-managed non-potable reuse might be considered 'light green' technologies, which reduce impacts on the environment but do not encourage wider ethical transformation of resource-using and resource-sharing practices. If users are able to continue to use water as a commodity resource provided by a centralised utility, then they may continue to be disconnected from the ecological and hydrological realities of their local environment.

Reuse and sustainable cities

Water reuse is a broad category for a range of strategies and technologies that have an obvious part to play in sustainable cities. Potable water reuse can help alleviate water supply deficits and reduce wastewater discharge to the environment with minimal change to water use and infrastructure, but at high cost and energy intensity, and with complex issues of social acceptability. Neighbourhood-scale non-potable reuse through dual-reticulation of water of varying qualities also provides relief to local water resources but at high cost, energy and resource intensity. Smaller-scale greywater reuse and rainwater harvesting can be part of green building design and may represent an emerging environmental ethic by building in technologies that acknowledge the limits to human abstraction of water from local catchments. Small-scale water reuse may also reinforce individualistic responses to resource scarcity, enabling households or building owners with sufficient capital to maintain high water-using lifestyles independently of shared infrastructure and resources.

Water reuse is consistent with sustainable cities discourse relating to urban metabolism and the circular economy. However, the sustainability of water reuse depends upon the scale, the technology, and the cultural and institutional arrangements. Water reuse may reduce abstraction of water from local catchments, but at the cost of increasing urban or building energy consumption. The embodied energy and materials in reuse systems may further undermine the sustainability of cities and efforts to reduce overall consumption. Water reuse may also support local urban metabolism by providing a resource to irrigate urban and peri-urban farms and industry.

A key challenge for water reuse is managing the public health risk. In centralised reuse schemes, both potable and non-potable, risk-averse strategies to improve public confidence and mitigate misuse have led to high specifications for treatment, with high energy, operating and capital costs. This has undermined the feasibility of non-potable dual-reticulation schemes in particular, which have proven to be too costly to operate or replicate, despite the water saving achieved. Regulations which require high-specification treatment systems for domestic or building-scale reuse also reduce health risk but increase cost and energy use. Protecting public health has been a key function of modern urban water infrastructure, but specifications for high-quality water reuse and the associated energy and material intensity highlight the trade-off between risk avoidance and sustainability.

Reuse both reinforces and challenges conventional arrangements for urban water infrastructure. Centralised reuse represents minimal change to infrastructure, yet public concerns about potable reuse demonstrate the importance of widening the range of interests represented in water resource and infrastructure decisions. Decentralised, low-energy systems that are managed by individuals or community-based interests represent a more radical departure from centralised infrastructure provision, but this can take the form of liberal self-reliance or an environmental ethic of shared concern about the local environment. The technologies and institutions of water reuse reflect wider debates and competing visions for urban sustainability – from technological innovation improving the environmental performance of existing infrastructure to more decentralised, communal approaches to managing resources within ecological and cultural limits.

References

Aitken, V., Bell, S., Hills, S. and Rees, L. 2014. Public Acceptability of Indirect Potable Water Reuse in the South-East of England. *Water Science and Technology: Water Supply* 14, 875–885. doi:10.2166/ws.2014.051

Angelakis, A.N. and Durham, B. 2008. Water Recycling and Reuse in EUREAU Countries: Trends and Challenges. *Desalination, AQUAREC 2006CHEMECA 2006* 218, 3–12. doi:10.1016/j.desal.2006.07.015

Asano, T. and Levine, A.D. 1996. Wastewater Reclamation, Recycling and Reuse: Past, Present, and Future. *Water Science and Technology* 33, 1–14.

Bakker, K. 2010. *Privatizing Water*. Cornell University Press, Ithaca and London.

Barton, A.B. and Argue, J.R. 2009. Integrated Urban Water Management for Residential Areas: A Reuse Model. *Water Science and Technology* 60(3), 813–823.

Bell, S. and Aitken, V. 2008. The Socio-Technology of Indirect Potable Water Reuse. *Water Science & Technology: Water Supply* 8, 441. doi:10.2166/ws.2008.104

Bixio, D., Thoeye, C., Wintgens, T., Ravazzini, A., Miska, V., Muston, M., Chikurel, H., Aharoni, A., Joksimovic, D. and Melin, T. 2008. Water Reclamation and Reuse: Implementation and Management Issues. *Desalination* 218, 13–23. doi:10.1016/j.desal.2006.10.039

Bridgeman, J. 2004. Public Perception Towards Water Recycling in California. *Water and Environment Journal* 18, 150–154. doi:10.1111/j.1747–6593.2004. tb00517.x

Brown, R., Farrelly, M. and Keath, N. 2009. Practitioner Perceptions of Social and Institutional Barriers to Advancing a Diverse Water Source Approach in Australia. *International Journal of Water Resources Development* 25, 15–28. doi:10.1080/07900620802586090

BSI. 2010. Greywater Systems – Part 1: Code of Practice, BS 8525–8521:2010.

Campisano, A., Butler, D., Ward, S., Burns, M.J., Friedler, E., DeBusk, K., Fisher-Jeffes, L.N., Ghisi, E., Rahman, A., Furumai, H. and Han, M. 2017. Urban

Rainwater Harvesting Systems: Research, Implementation and Future Perspectives. *Water Research* 115, 195–209. doi:10.1016/j.watres.2017.02.056

Castán Broto, V., Allen, A. and Rapoport, E. 2012. Interdisciplinary Perspectives on Urban Metabolism. *Journal of Industrial Ecology* 16, 851–861. doi:10.1111/j.1530–9290.2012.00556.x

Colebatch, H.K. 2006. Governing the Use of Water: The Institutional Context. *Desalination* 187, 17–27. doi:10.1016/j.desal.2005.04.064

Cooley, H. and Wilkinson, R. 2012. *Implications of Future Water Supply Sources for Energy Demands*. Water Reuse Association, Alexandria.

Cooper, E. 2003. Rouse Hill and Picton Reuse Schemes: Innovative Approaches to Large-Scale Reuse. *Water, Science and Technology: Water Supply* 3, 49–54.

Crettaz, P., Jolliet, O., Cuanillon, J-M. and Orlando, S. 1999. Life Cycle Assessment of Drinking Water and Rain Water for Toilets Flushing. *Journal of Water Supply: Research and Technology – Aqua* 48, 73–83.

Department of Health. 2010. *Code of Practice for the Reuse of Greywater in Western Australia 2010*. Western Australian Government, Perth.

Dolnicar, S., Hurlimann, A. and Grün, B. 2011. What Affects Public Acceptance of Recycled and Desalinated Water? *Water Research* 45, 933–943. doi:10.1016/j.watres.2010.09.030

du Pisani, P.L. 2006. Direct Reclamation of Potable Water at Windhoek's Goreangab Reclamation Plant. *Desalination, Integrated Concepts in Water Recycling* 188, 79–88. doi:10.1016/j.desal.2005.04.104

Ensink, J.H.J., Hoek, W. van der, Matsuno, Y., Munir, S. and Aslam, M.R. 2002. *Use of Untreated Wastewater in Peri-Urban Agriculture in Pakistan: Risks and Opportunities*. IWMI, Colombo.

Environment Agency. 2010. *Rainwater Harvesting for Domestic Uses: An Information Guide*. Environment Agency, Bristol.

Environment Agency. 2011. *Greywater for Domestic Uses: An Information Guide*. Environment Agency, Bristol.

Ferguson, B.C., Frantzeskaki, N. and Brown, R.R. 2013. A Strategic Program for Transitioning to a Water Sensitive City. *Landscape and Urban Planning* 117, 32–45. doi:10.1016/j.landurbplan.2013.04.016

Friedler, E. 2001. Water Reuse – An Integral Part of Water Resources Management. *Water Policy* 3, 29–39. doi:10.1016/S1366–7017(01)00003–00004

Geels, F.W. 2002. Technological Transitions as Evolutionary Reconfiguration Processes: A Multi-Level Perspective and a Case-Study. *Research Policy, NELSON + WINTER + 20* 31, 1257–1274. doi:10.1016/S0048–7333(02)00062–00068

Gikas, P. and Tchobanoglous, G. 2009. The Role of Satellite and Decentralized Strategies in Water Resources Management. *Journal of Environmental Management* 90, 144–152. doi:10.1016/j.jenvman.2007.08.016

Guo, T. and Englehardt, J.D. 2015. Principles for Scaling of Distributed Direct Potable Water Reuse Systems: A Modeling Study. *Water Research* 75, 146–163. doi:10.1016/j.watres.2015.02.033

Handia, L., Tembo, J.M. and Mwiindwa, C. 2003. Potential of Rainwater Harvesting in Urban Zambia. *Physics and Chemistry of the Earth* Parts A/B/C 28, 893–896. doi:10.1016/j.pce.2003.08.016

Hartley, T.W. 2006. Public Perception and Participation in Water Reuse. *Desalination* 187, 115–126. doi:10.1016/j.desal.2005.04.072

Hatt, B.E., Deletic, A. and Fletcher, T.D. 2006. Integrated Treatment and Recycling of Stormwater: A Review of Australian Practice. *Journal of Environmental Management* 79, 102–113. doi:10.1016/j.jenvman.2005.06.003

Hermanowicz, S.W. and Asano, T. 1999. Abel Wolman's "The Metabolism of Cities" Revisited: A Case for Water Recycling and Reuse. *Water Science and Technology* 40, 29–36. doi:10.1016/S0273-1223(99)00482-00485

Huertas, E., Salgot, M., Hollender, J., Weber, S., Dott, W., Khan, S., Schäfer, A., Messalem, R., Bis, B., Aharoni, A. and Chikurel, H. 2008. Key Objectives for Water Reuse Concepts. *Desalination* 218, 120–131. doi:10.1016/j.desal.2006.09.032

Hurlimann, A. and Dolnicar, S. 2010. When Public Opposition Defeats Alternative Water Projects – The Case of Toowoomba Australia. *Water Research* 44, 287–297. doi:10.1016/j.watres.2009.09.020

Hurlimann, A.C. and McKay, J.M. 2006. What Attributes of Recycled Water Make It Fit for Residential Purposes? The Mawson Lakes Experience. *Desalination* 187, 167–177. doi:10.1016/j.desal.2005.04.077

Hwang, H., Forrester, A. and Lansey, K. 2014. *Decentralized Water Reuse: Regional Water Supply System Resilience Benefits*. Procedia Engineering, 12th International Conference on Computing and Control for the Water Industry, CCWI2013 70, 853–856. doi:10.1016/j.proeng.2014.02.093

Jefferson, B., Palmer, A., Jeffrey, P., Stuetz, R. and Judd, S. 2004. Grey Water Characterisation and Its Impact on the Selection and Operation of Technologies for Urban Reuse. *Water Science and Technology* 50, 157–164.

Joss, A., Siegrist, H. and Ternes, T.A. 2008. Are We About to Upgrade Wastewater Treatment for Removing Organic Micropollutants? *Water Science and Technology* 57, 251–255. doi:10.2166/wst.2008.825

Lahnsteiner, J., Rensburg, P. van and Esterhuizen, J. 2017. Direct Potable Reuse – A Feasible Water Management Option. *Journal of Water Reuse and Desalination* jwrd2017172. doi:10.2166/wrd.2017.172

Law, I.B. 1996. Rouse Hill – Australia's First Full Scale Domestic Non-Potable Reuse Application. *Water Science and Technology* 33, 71–78. doi:10.1016/0273-1223(96)00408-00408

Lazarova, V., Asano, T., Bahri, A. and Anderson, J. 2013. *Milestones in Water Reuse*. IWA Publishing, London.

Lazarova, V., Hills, S. and Birks, R. 2003. Using Recycled Water for Non-Potable, Urban Uses: A Review with Particular Reference to Toilet Flushing. *Water Science & Technology: Water Supply – WSTWS* 10, 66–77.

Leverenz, H.L., Tchobanoglous, G. and Asano, T. 2011. Direct Potable Reuse: A Future Imperative. *Journal of Water Reuse and Desalination* 1, 2–10. doi:10.2166/wrd.2011.000

Li, Z., Boyle, F. and Reynolds, A. 2010. Rainwater Harvesting and Greywater Treatment Systems for Domestic Application in Ireland. *Desalination* 260, 1–8. doi:10.1016/j.desal.2010.05.035

Lim, S-R., Suh, S., Kim, J-H. and Park, H.S. 2010. Urban Water Infrastructure Optimization to Reduce Environmental Impacts and Costs. *Journal of Environmental Management* 91, 630–637. doi:10.1016/j.jenvman.2009.09.026

Livingston, D., Stenekes, N., Colebatch, H., Ashbolt, N. and Waite, D. 2004. *Water Recycling and Decentralised Management: The Policy and Organisational Challenges for Innovative Approaches*. Presented at the WSUD 2004: Cities as

Catchments; International Conference on Water Sensitive Urban Design, Engineers Australia, Canberra, A.C.T.

Lundie, S., Peters, G.M. and Beavis, P.C. 2004. Life Cycle Assessment for Sustainable Metropolitan Water Systems Planning. *Environmental Science & Technology* 38, 3465–3473. doi:10.1021/es034206m

Mahesh, J., Amerasinghe, P. and Pavelic, P. 2015. An Integrated Approach to Assess the Dynamics of a Peri-Urban Watershed Influenced by Wastewater Irrigation. *Journal of Hydrology* 523, 427–440. doi:10.1016/j.jhydrol.2015.02.001

Makoni, F.S., Thekisoe, O.M.M. and Mbati, P.A. 2016. Urban Wastewater for Sustainable Urban Agriculture and Water Management in Developing Countries, in: *Sustainable Water Management in Urban Environments, The Handbook of Environmental Chemistry*. Springer, Cham, pp. 265–293. doi:10.1007/978-3-319-29337-0_9

Makropoulos, C.K. and Butler, D. 2010. Distributed Water Infrastructure for Sustainable Communities. *Water Resource Management* 24, 2795–2816. doi:10.1007/s11269-010-9580-5

Mandal, D., Labhasetwar, P., Dhone, S., Dubey, A.S., Shinde, G. and Wate, S. 2011. Water Conservation Due to Greywater Treatment and Reuse in Urban Setting with Specific Context to Developing Countries. *Resources, Conservation and Recycling* 55, 356–361. doi:10.1016/j.resconrec.2010.11.001

Mankad, A. 2012. Decentralised Water Systems: Emotional Influences on Resource Decision Making. *Environment International* 44, 128–140. doi:10.1016/j.envint.2012.01.002

Marks, J., Martin, B. and Zadoroznyj, M. 2008. How Australians Order Acceptance of Recycled Water National Baseline Data. *Journal of Sociology* 44, 83–99. doi:10.1177/1440783307085844

Marks, J.S. 2006. Taking the Public Seriously: The Case of Potable and Non Potable Reuse. *Desalination* 187, 137–147. doi:10.1016/j.desal.2005.04.074

McCarty, P.L., Bae, J. and Kim, J. 2011. Domestic Wastewater Treatment as a Net Energy Producer – Can This Be Achieved? *Environmental Science & Technology* 45, 7100–7106. doi:10.1021/es2014264

McClelland, C.J., Linden, K., Drewes, J.E., Khan, S.J., Raucher, R. and Smith, J. 2012. Determining Key Factors and Challenges That Affect the Future of Water Reuse. *Journal of Water Supply Research & Technology-Aqua* 61, 518–528. doi:10.2166/aqua.2012.188

Melin, T., Jefferson, B., Bixio, D., Thoeye, C., De Wilde, W., De Koning, J., van der Graaf, J. and Wintgens, T. 2006. Membrane Bioreactor Technology for Wastewater Treatment and Reuse. *Desalination, Integrated Concepts in Water Recycling* 187, 271–282. doi:10.1016/j.desal.2005.04.086

Memon, F.A., Makropoulos, C., Sakellari, I. and Butler, D. 2005. Modelling Sustainable Urban Water Management Options. *Proceedings of the ICE – Engineering Sustainability* 158, 143–153. doi:10.1680/ensu.2005.158.3.143

Miller-Robbie, L., Ramaswami, A. and Amerasinghe, P. 2017. Wastewater Treatment and Reuse in Urban Agriculture: Exploring the Food, Energy, Water, and Health Nexus in Hyderabad, India. *Environmental Research Letters* 12, 075005. doi:10.1088/1748-9326/aa6bfe

Mills, W.R. and Watson, I.C. 1994. Water Factory 21 – The Logical Sequence. *Desalination* 98, 265–272. doi:10.1016/0011-9164(94)00151-00150

Mitchell, V. 2006. Applying Integrated Urban Water Management Concepts: A Review of Australian Experience. *Environmental Management* 37, 589–605. doi:10.1007/s00267-004-0252-1

Mitchell, V.G., Mein, R.G. and McMahon, T.A. 2002. Utilising Stormwater and Wastewater Resources in Urban Areas. *Australasian Journal of Water Resources* 6, 31–43. doi:10.1080/13241583.2002.11465208

Moglia, M., Gan, K., Delbridge, N., Sharma, A.K. and Tjandraatmadja, G. 2016. Investigation of Pump and Pump Switch Failures in Rainwater Harvesting Systems. *Journal of Hydrology* 538, 208–215. doi:10.1016/j.jhydrol.2016.04.020

Okun, D.A. 1969. Alternatives in Water Supply. *Journal of the American Water Works Association* 61(5), 215–224.

Okun, D.A. 1973. Planning for Water Reuse. *Journal (American Water Works Association)* 65, 617–622.

Okun, D.A. 1997. Distributing Reclaimed Water Through Dual Systems. *Journal (American Water Works Association)* 89, 52–64.

Parkes, C., Kershaw, H., Hart, J., Sibille, R. and Grant, Z. 2010. *Energy and Carbon Implications of Rainwater Harvesting and Greywater Recycling*. Final Report. Environment Agency, Bristol.

Price, J., Fielding, K. and Leviston, Z. 2012. Supporters and Opponents of Potable Recycled Water: Culture and Cognition in the Toowoomba Referendum. *Society & Natural Resources* 25, 980–995. doi:10.1080/08941920.2012.656185

Qadir, M., Wichelns, D., Raschid-Sally, L., McCornick, P.G., Drechsel, P., Bahri, A. and Minhas, P.S. 2010. The Challenges of Wastewater Irrigation in Developing Countries. *Agricultural Water Management, Comprehensive Assessment of Water Management in Agriculture* 97, 561–568. doi:10.1016/j.agwat.2008.11.004

Radcliffe, J. 2015. Water Recycling in Australia – During and After the Drought. *Environmental Science: Water Research & Technology* 1, 554–562. doi:10.1039/C5EW00048C

Russell, S. and Lux, C. 2009. Getting Over Yuck: Moving from Psychological to Cultural and Sociotechnical Analyses of Responses to Water Recycling. *Water Policy* 11, 21. doi:10.2166/wp.2009.007

Sapkota, M., Arora, M., Malano, H., Moglia, M., Sharma, A., George, B. and Pamminger, F. 2014. An Overview of Hybrid Water Supply Systems in the Context of Urban Water Management: Challenges and Opportunities. *Water* 7, 153–174. doi:10.3390/w7010153

Sofoulis, Z. 2005. Big Water, Everyday Water: A Sociotechnical Perspective. *Continuum: Journal of Media & Cultural Studies* 19, 445–463. doi:10.1080/10304310500322685

Srinivasan, V., Gorelick, S.M. and Goulder, L. 2010. Sustainable Urban Water Supply in South India: Desalination, Efficiency Improvement, or Rainwater Harvesting? *Water Resources Research* 46, W10504. doi:10.1029/2009WR008698

Stokes, J. and Horvath, A. 2006. Life Cycle Energy Assessment of Alternative Water Supply Systems (9 pp). *The International Journal of Life Cycle Assessment* 11, 335–343. doi:10.1065/lca2005.06.214

Sturgis, P. and Allum, N. 2004. Science in Society: Re-Evaluating the Deficit Model of Public Attitudes. *Public Understanding of Science* 13, 55–74. doi:10.1177/0963662504042690

Tangsubkul, N., Beavis, P., Moore, S., Lundie, S. and Waite, T. 2005. Life Cycle Assessment of Water Recycling Technology. *Water Resources Management* 19, 521–537. doi:10.1007/s11269-005-5602-0

Tjandraatmadja, G., Burn, S., McLaughlin, M. and Biswas, T. 2005. Rethinking Urban Water Systems – Revisiting Concepts in Urban Wastewater Collection and Treatment to Ensure Infrastructure Sustainability. *Water Science and Technology: Water Supply* 5, 145–154.

Toze, S. 2006. Water Reuse and Health Risks – Real vs. Perceived. *Desalination* 187, 41–51. doi:10.1016/j.desal.2005.04.066

Venkatesh, G. and Brattebø, H. 2011. Energy Consumption, Costs and Environmental Impacts for Urban Water Cycle Services: Case Study of Oslo (Norway). *Energy* 36, 792–800. doi:10.1016/j.energy.2010.12.040

Verrecht, B., James, C., Germain, E., Birks, R., Barugh, A., Pearce, P. and Judd, S. 2012. Economical Evaluation and Operating Experiences of a Small-Scale MBR for Nonpotable Reuse. *Journal of Environmental Engineering* 138, 594–600. doi:10.1061/(ASCE)EE.1943–7870.0000505

WHO. 2006. *Guidelines for the Safe Use of Wastewater, Excreta and Greywater.* World Health Organization, Geneva.

Wilcox, J., Nasiri, F., Bell, S. and Rahaman, M.S. 2016. Urban Water Reuse: A Triple Bottom Line Assessment Framework and Review. *Sustainable Cities and Society* 27, 448–456. doi:10.1016/j.scs.2016.06.021

Willis, R.M., Stewart, R.A., Williams, P.R., Hacker, C.H., Emmonds, S.C. and Capati, G. 2011. Residential Potable and Recycled Water End Uses in a Dual-Reticulated Supply System. *Desalination* 272, 201–211. doi:10.1016/j.desal.2011.01.022

Wisniewski, C. 2007. Membrane Bioreactor for Water Reuse. *Desalination, EuroMed* 2006 203, 15–19. doi:10.1016/j.desal.2006.05.002

Wong, T.H.F. and Brown, R.R. 2009. The Water Sensitive City: Principles for Practice. *Water Science and Technology* 60, 673–682. doi:10.2166/wst.2009.436

9 Desalination

Introduction

Water covers 70% of the earth's surface, but 97.5% of it is salty and less than 0.01% is freshwater available in rivers, reservoirs and lakes (Shiklomanov, 2000). Turning salty water into fresh seems an obvious answer to water scarcity. Desalination appears to be a straightforward technical solution that can provide limitless supply of water for human use, yet it has been highly controversial. Significant concerns about high costs, energy requirements and environmental impacts undermine the promise of desalination as a limitless new source of freshwater for cities.

Producing drinking water from saline sources requires a lot of energy. Desalination has traditionally been based on distillation, but membrane technologies have dominated new desalination construction since the 1990s (Ziolkowska, 2016). The energy intensity of water treatment increases with the level of contaminants in the raw water source (Table 9.1). Desalination is the most energy intensive of water treatment methods, as saline and brackish waters have higher contaminant concentrations than recycled or conventional water resources. In 2013 around 60% of desalination plants installed globally were designed to treat seawater, 22% to treat brackish water sources, and the remainder to treat water from rivers and wastewater (Burn et al., 2015).

In most applications, desalination is more expensive than other water treatment technologies (Ghaffour et al., 2013). High energy requirements contribute to high operating costs, and complex technologies mean high capital costs. Technical developments in membranes and system design have reduced energy requirements and operating and capital costs in recent years, but they remain high compared to conventional supply (Ziolkowska, 2015). In areas with severe scarcity of freshwater, desalination can be cost effective compared to alternatives such as long-distance water transfer schemes, but in most cities desalination remains the most expensive and energy-intensive option for supplying water.

Desalination supplies less than 1% of the world's water use, but it has been expanding rapidly in recent years and has come to dominate supplies in a

Table 9.1 Energy intensity of water treatment technologies (Cooley and Wilkinson, 2012)

Treatment technology	Energy intensity (kWh/ML)
Conventional drinking water treatment	32–530
Recycled water using membranes	845–2,200
Brackish water desalination using membranes	790–2,200
Seawater desalination using membranes	>3,100

few water-scarce cities. In 2013 60% of desalinated water was for municipal supply, 28% for industry, 6% for electricity production and the remainder for agriculture, tourism and other uses (Burn et al., 2015). In 2015 there were 18,426 desalination plants installed in 150 countries worldwide (International Desalination Association, 2017). Desalination supplies 100% of water in Qatar and Kuwait, 55% in Israel and 50% in the Australian city of Perth (Ghaffour, 2009; Jacobsen, 2016; Water Corporation, 2017). Desalination more commonly forms a relatively small proportion of total supply and may be installed as a drought resilience measure or emergency response.

Growth in desalination is attributed to rising demand for water and falling capital and operating costs. Global capacity grew at around 8.5% per annum in the first decade of the twenty-first century (Boals, 2009; Virgili and Pankratz, 2016). Most growth has been in the Middle East, which has long-established desalination experience and in 2015 held 44% of the world's desalination capacity (Voutchkov, 2016). Desalination capacity also increased significantly in China, Australia, Spain, the US, Japan and other countries. particularly in response to drought (Ziolkowska, 2016). Increasing desalination capacity in China is mostly associated with demand for freshwater from the industrial sector, particularly for cooling water for both thermal and nuclear power plants (Avrin et al., 2015; Zheng et al., 2014). In Spain desalination plants have been installed to meet agricultural as well as domestic demand, as an alternative to inter-basic water transfers (March, 2015; Swyngedouw, 2013).

In 2015 investment in new desalination plants was US$21M and was forecast to double by 2020 (Runte, 2016). However, forecasts for growth to continue to increase between 2010 and 2020 have not been fulfilled, with a peak in new contracts for desalination in 2007 not exceeded by 2016 (Virgili and Pankratz, 2016). The slowdown in growth in desalination can be attributed to general economic slowdown and global instability since 2008, but also reflects a more cautious approach to investment after the initial rapid expansion of desalination at the beginning of the century.

Installation of desalination plants that have never operated, such as in Sydney, San Diego and London, demonstrate trade-offs between financial investment in an expensive source of water and drought risk management (Porter et al., 2015; Scarborough et al., 2015). These can be considered 'stranded assets', leaving water utilities with long-term contracts and capital costs for a desalination plant that is not needed or too expensive to operate (Desalination and Water Reuse, 2015; Turner et al., 2016). The certainty of supply from desalination can be attractive to decision-makers in the midst of the uncertainty of drought, but may not be the best short- or long-term economic or environmental option for alleviating supply-demand deficits.

For some proponents, policy-makers and researchers, renewably powered desalination is the ultimate sustainable solution to water scarcity, and it is only a matter of time and continued innovation before it is economically competitive with conventional water sources. This is attributed to rapid reductions in cost and to the growing problem of water scarcity and drought. Concerns about carbon emissions have been addressed in the UK, the US and Australia by building renewable energy schemes to power new desalination plants. However, according to its critics, desalination is an essentially unsustainable technology which continues patterns of development, resource exploitation and pollution that have contributed to the current environmental crisis.

This chapter provides an overview of the key technologies that constitute desalination in different parts of the world, including prospects for future innovation. It analyses desalination through different framings of urban water sustainability discourses. The role of desalination in urban water systems is highly dependent on local conditions and politics and connects to wider global trends in technology, economics and development.

Desalination methods

There are two basic mechanisms for desalination – thermal and membrane separation. Thermal desalination involves evaporating the water, leaving a more concentrated solution behind, and condensing the water vapour to obtain freshwater (Gebel, 2014). Modern thermal desalination processes include multi-stage flash (MSF) distillation and multi-effect distillation (MED). The principle of membrane separation is that water passes through a semi-permeable membrane but salt ions do not, creating freshwater on one side of the membrane and a more concentrated salt solution on the other. The most prevalent membrane technology for desalination is reverse osmosis (RO). Thermal desalination was the dominant process in global installations until the 1990s, when reverse osmosis became more reliable, energy efficient and affordable. By 2013, more than 70% of global desalinated water was treated using RO (Ziolkowska, 2015).

Thermal desalination

Aristotle observed the purification of salt water by evaporation in ancient Greece, and distillation has been used since the sixteenth century to provide freshwater on ships during long journeys (Birkett, 2010; Kumar et al.,

2017). Early industrial technologies for desalination were simple distillation units, evaporating seawater in a still, leaving behind a concentrated brine, and producing freshwater by condensing the steam in a water-cooled condenser. Energy demands for boiling the water meant that desalination was only attractive in isolated circumstances, particularly for long-distance ocean voyages and for the military (Belessiotis and Delyannis, 2000; Delyannis, 2003). Desalination technologies were the subject of several patents during this period, and scientific and technical debates addressed the efficacy of various additives, the fuel required and the best configuration of the distillation equipment for routine or emergency use (see for example, Lind, 1811).

The first major improvement in the efficiency of desalination was the development of multi-effect distillation (MED). MED was initially developed for sugar refining, and it uses a series of connected evaporation cells, each at a lower temperature than the previous one. In each cell, seawater is sprayed over a pipe containing steam, causing water to condense within the pipe. The seawater evaporates in the cell, producing steam that is transferred into the next chamber in a closed pipe, where it condenses due to the lower temperature (Figure 9.1).

The next major development in thermal desalination processes was multi-stage flash (MSF) distillation, developed in the 1930s (Figure 9.2). MSF also consists of a series of chambers of decreasing temperature and pressure. In this system seawater flows through the tubes of a heat exchanger, from the coolest to the hottest chamber, preheating it to the highest temperature. The heated seawater then flows into the brine reservoir of the hottest chamber. Water evaporates from the brine and condenses on the heat exchanger pipes at the top of the chamber. The condensate is collected to provide the

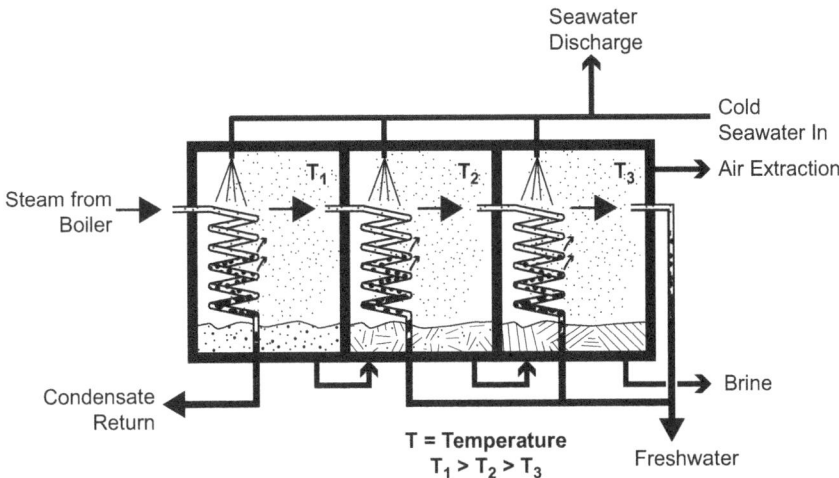

Figure 9.1 Multi-effect distillation desalination

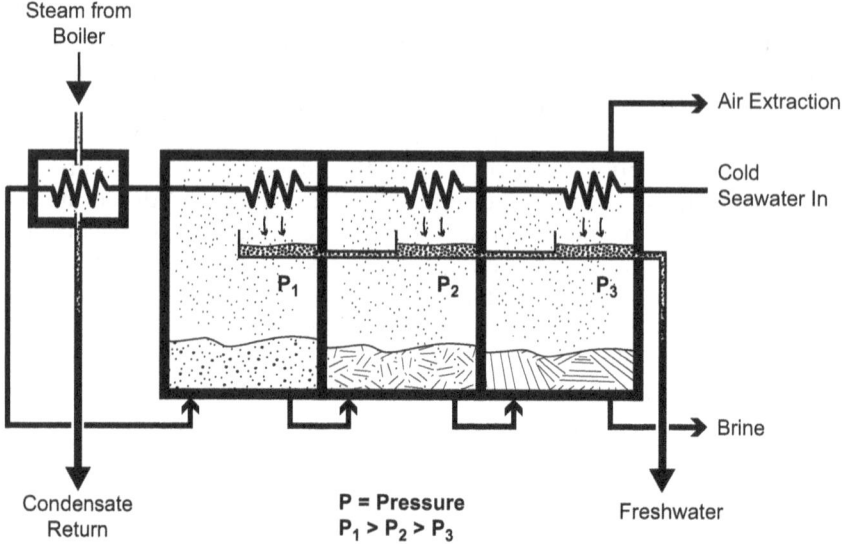

Steam from Boiler

Air Extraction

Cold Seawater In

P_1

P_2

P_3

Brine

Condensate Return

P = Pressure
$P_1 > P_2 > P_3$

Freshwater

Figure 9.2 Multi-stage flash desalination

freshwater product. Brine is pumped from the hot high-pressure chamber to the next chamber, causing a 'flash' of evaporation as hot water encounters cooler, lower-pressure conditions. This vapour also condenses on the heat exchanger, producing freshwater, which is collected and finally discharged from the coolest chamber. This process of heating and 'flashing' continues until the final stage, where the concentrated, cooled brine is discharged.

Vapour compression distillation (VCD) is the third-most common thermal desalination method and is used on its own or in combination with MED and MSF processes (Aly and El-Figi, 2003; El-Dessouky et al., 2000). Independent VCD systems are run on electricity, do not require a heat source, are relatively compact and are most applicable to small-scale applications in remote locations (Ettouney et al., 1999). Water vapour from the salt water is drawn into a mechanical compressor. Compression of the vapour leads to condensation of freshwater and releases latent heat, which is used to heat the salt water intake to increase evaporation.

Modern thermal desalination methods have been used since the 1930s and are still widely in operation, particularly in the Middle East. Thermal processes are attractive in countries with low fuel costs, particularly when desalination is collocated with thermal power plants, utilising waste heat and steam. Middle Eastern countries with access to oil but limited freshwater resources were early investors in thermal desalination plants. Thermal desalination has also been associated with nuclear power plants, and the concept of 'nuclear desalination' was the focus of research and development in the 1950s and '60s. Nuclear desalination has been implemented in Japan, India and Kazakhstan, using both thermal and membrane processes.

Membrane desalination

Interest in desalination technologies increased rapidly after the Second World War. The formation of the Office of Saline Water in the US Department of the Interior led to significant research and development of desalination technologies during the 1950s and '60s (MacGowan, 1963). This included investment in basic research and early trials into the use of membranes for desalination.

Electrodialysis was the first membrane-based desalination process to be applied in practice. It was developed in the 1950s and is most suitable for small-scale desalination of brackish water (Lee and Moon, 2014). Electrodialysis involves applying a current to a solution to draw dissolved salts towards positively and negatively charged electrodes. Ion exchange membranes separate concentrated salt solution near the electrodes from the main solution. As the water passes through a series of electrodialysis cells the main solution becomes increasingly fresh, as more salt ions are drawn through the ion exchange membranes to the electrodes.

Since the 1950s the main developments in desalination have been in membrane separation technologies. Research undertaken in the Department of Chemistry at the University of Florida identified the potential for cellulose acetate to act as a semi-impermeable membrane for reverse osmosis desalination (Reid and Breton, 1959). Simultaneously, researchers at UCLA were testing similar membranes and developed their own formulation for asymmetric membranes, building on the work of French researcher Dobry from the 1930s (Dobry, 1936 in Glater, 1998; Loeb, 1981; Loeb and Sourirajan, 1963). The first large-scale seawater desalination plant was constructed in Jedda, Saudi Arabia, in 1975 (Al-Gholaikah et al., 1978). Early RO membranes had problems with breaking, fouling and a low flux of water due to low permeability. In the 1980s thin film composite materials enabled the production of polyamide RO membranes with higher permeability and durability, paving the way for more efficient and affordable membranes that have enabled the recent rapid expansion in global desalination capacity (Cadotte, 1981; Kumar et al., 2017).

In water supply applications, reverse osmosis membranes are typically wound in several layers around a pipe, contained within an outer cylindrical casing (Kumar et al., 2017; Wilf, 2014). Salt water is pumped under pressure through the outer layers of membrane from one end of the cylindrical module, freshwater is collected in the inner pipe, and concentrated brine is discharged at the other end of the module. A utility-scale desalination plant typically consists of hundreds of membrane modules, arranged in blocks or banks (Figure 9.3). Desalination may occur in two or more steps, with one set of membranes producing water of an intermediate salinity, which is

Figure 9.3 Reverse osmosis desalination plant

treated a second time to achieve the required final water quality. Membranes are also regularly back flushed to remove scale and impurities, requiring sufficient membrane capacity to maintain production, allowing for cleaning and maintenance. The membranes themselves are just one process step in a desalination plant, which also includes pre-treatment to remove suspended solids, chemical dosing to reduce scale and extend membrane life, and post-treatment such as chlorination and dosing with minerals to ensure safe distribution of the water into supply.

Improving RO membrane performance has increased permeability and reduced the pressure required to produce freshwater. RO membranes can now operate at around 1.1–1.2 times the osmotic pressure of seawater (Elimelech and Phillip, 2011). Further reductions in operating pressure are limited by thermodynamics and cannot go below the osmotic pressure of the concentration of the intake flow. Innovation in membrane design, including the use of carbon nanotubes, aquaporin proteins to mimic natural membranes, and graphene-based membranes, can improve the flow rate across the membrane but cannot reduce the pressure required below this thermodynamic limit (Surwade et al., 2015; Tang et al., 2013). This can reduce the size of membrane required to achieve production rates, reducing capital costs, but operating pressures will remain above the osmotic pressure of the salt water feed (Elimelech and Phillip, 2011; Ghaffour et al., 2013). As membranes come close to the thermodynamic limit of operating pressure, improving the design and efficiency of other processes and systems is of increasing importance in reducing costs (Elimelech and Phillip, 2011; Wilf, 1997). The operating cost of RO desalination has been reduced significantly through the implementation of energy recovery, with recovery efficiency now up to 98% (Busch and Mickols, 2004; Ghaffour et al., 2013).

Environment and energy

The most significant local environmental impacts of desalination plants are due to the intake of water from the environment and the discharge of concentrated brine, which may also be contaminated with chemicals used in pre-treatment (Miri and Chouikhi, 2005). For seawater desalination, intake design is increasingly important in reducing impacts on the marine environment, in particular the entrainment of fish and other marine life in the flow of seawater as it is pumped to the desalination plant. Screening of intakes can reduce entrainment of larger fish, but fish may become trapped against the screens and smaller fish and larvae can still be caught in the flow. Intakes designed to draw water from below the seabed, effectively drawing saline groundwater, reduce these impacts but are more expensive to build and operate, requiring higher pumping pressure to abstract the seawater (Lattemann and Höpner, 2008; Mackey et al., 2011).

Discharge of concentrated brine into the environment is another potential impact of desalination (Gilron et al., 2003). This is particularly an issue for inland brackish water desalination, where the brine may be of substantially higher salinity than the receiving water, which may also be locally contained (Ahmed et al., 2001). In such cases discharging brine back into the environment could increase the salinity of the brackish water body, and brine may instead be evaporated to produce salts, which may be of commercial value. Discharging brine into the ocean may increase salinity and change water temperature close to the discharge point, with localised impacts on marine or estuarine ecosystems. Chemicals used in pre-treatment may also be discharged with the brine, increasing the risk of specific environmental impacts. Design and placement of discharge is therefore important to minimise environmental harm, and desalination may not be appropriate in areas of high ecological significance or vulnerability.

Reducing energy use has been the most important factor in reducing operating costs, and it is also important in reducing carbon emissions of desalination. Improved efficiency of both thermal and membrane processes has led to significant reductions in energy demands for desalination, but this cannot go below thermodynamic limits (Elimelech and Phillip, 2011; Ghaffour et al., 2013; Subramani and Jacangelo, 2015).

Most desalination is currently powered using fossil fuels (Ghaffour et al., 2015). At the 2015 Paris Climate Change Conference an alliance was formed to set targets for the reduction of carbon emissions associated with desalination. The Global Clean Water Desalination Alliance proposed that by 2035 80% of new desalination and 10% of the total should be powered by renewables (Global Clean Water Desalination Alliance, 2015). This industry initiative is non-binding, but it demonstrates awareness of the challenge that carbon emissions present to the expansion of desalination.

Renewable energy has been used to meet the energy demands of desalination in Australia, the UK and the US. Renewable energies can be directly used to power desalination, as in solar thermal and geothermal processes (Abutayeh et al., 2014; Delyannis, 2003). In the case of RO and other requirements for electricity, it is more effective to have renewable energy sources connected to the grid, to allow for variability in energy production. The Gateway desalination plant at Beckton in London is nominally powered from energy generated from methane produced from the nearby wastewater treatment plant and renewable energy purchased from the grid. The Perth desalination plant in Western Australia is also powered by a wind farm and solar array 400km to the north near Geraldton (Water Corporation, 2017). Solar photovoltaics can be used in small-scale brackish RO desalination for remote applications.

Isolated desalination plants such as those in London and Sydney have addressed sustainability concerns by powering their systems with renewable energy. However, in the broader industrial context, using renewable energy to meet unprecedented demands from energy-intensive desalination plants displaces urgently needed reductions in overall carbon emissions. The UK water industry uses roughly 3% of generated electricity and in the US water-related energy use is responsible for nearly 5% of all greenhouse gas emissions (Rothausen and Conway, 2011). In the transition to a sustainable low-carbon economy, water utilities, like all industries, is challenged to reduce total energy use and meet existing energy demand with renewable energy, rather than simply using renewable energy to justify increased energy consumption.

Nuclear desalination has been proposed as a technical solution to overcome both water and energy limits (Megahed, 2003). Nuclear-powered aircraft carriers use desalination in the US Navy. Co-locating nuclear power stations with thermal desalination, using seawater as cooling water and low-pressure steam from the turbine, has been proposed, as has using electricity from a nuclear power plant for RO (Megahed, 2001). Desalination is seen as a potential means of balancing load for nuclear power stations, using electricity to desalinate water at times when the demand for electricity from the grid is low. Nuclear desalination has been promoted since the 1960s and has been used in Japan, Kazakhstan and India. In Japan MSF, MED and RO desalination have been coupled with ten nuclear power reactors, and the water produced is used in the power stations' own cooling systems. A nuclear reactor in Aktau, Kazakhstan, used MED to supply water to the city between 1973 and 1999. MVC and RO desalination are used at the Kundankulam nuclear plant near Chennai in India, mostly to supply cooling water to the plant and to provide drinking water to a nearby town. Nuclear desalination plants have also been investigated and proposed in China, Pakistan, Russia and the United States.

Framing desalination

Proponents claim that desalination is a sustainable solution to water crises, particularly when powered by renewable energy. Critics maintain that the high energy, high cost and environmental impacts mean that desalination is fundamentally unsustainable and that it diverts efforts away from more sustainable options such as demand management and reuse. In analysing the sustainability of desalination it is therefore important to understand how these different viewpoints frame urban water sustainability and the role of desalination as a specific technical option for cities.

Sustainable development

Proponents of desalination commonly refer to projections for the increasing number of people living under conditions of water stress as evidence of the growing global demand for this technology (see for example Elimelech and Phillip, 2011; Gude, 2016; International Desalination Association, 2017; Subramani and Jacangelo, 2015; Ziolkowska, 2016). Desalination presents a technical solution to the sustainable development challenge of improving standards of living for a growing global population within resource constraints. Desalination potentially provides technology to overcome constraints on water resources to deliver clean water to people living in water-scarce conditions (see Box 9.1).

As a technical solution, desalination avoids drawing attention to complex social, institutional, economic and political factors that contribute to water scarcity and lack of access to safe drinking water (Zeitoun et al., 2016). Water poverty is more often a result of lack of infrastructure than absolute water scarcity, and the high cost of desalination means it is unlikely to be the most affordable or preferable option in most cities in the Global South (Bakker, 2010; Cook and Bakker, 2012). Water scarcity also arises in part from over-exploitation of water resources for agriculture and industry, which can often be managed through better resource allocation policy and by encouraging higher efficiency rather than building desalination plants.

Small-scale reverse osmosis, with renewable energy, can provide clean water to remote communities, but a high level of technical expertise is required to monitor performance and maintain the systems. Middle-class and wealthy households may be able to afford to install their own RO systems to treat groundwater and pay for energy supply and maintenance where centralised water supply is intermittent and groundwater is brackish or saline. This can increase inequality of service within the city and undermine local groundwater reserves.

Box 9.1   Desalination in Chennai

In Chennai in India, desalination has been promoted as a sustainable solution to local water shortages and an alternative to unsustainable exploitation of groundwater by residents, industry and the water utility (Arabindoo, 2011; Srinivasan et al., 2010; Vedachalam, 2012). Over-exploitation of groundwater has led to saline intrusion, polluting the remaining fresh groundwater. Household-scale RO has become popular with middle-class and wealthy homeowners who have drilled their own wells. Centralised seawater desalination has been promoted and built as a sustainable solution to local resource exploitation and the comparative inefficiency of small-scale systems. Desalination has also been used to provide cooling water for a new power plant, as an alternative to exploiting local freshwater resources. With highly constrained freshwater resources, desalination has been presented as enabling sustainable development in Chennai, including development of the economy through electricity provision and reducing impacts on local water resources. However, it has also been subject to criticism, particularly the high cost, environmental impacts, energy demands and lost opportunities for improved demand management and better integration of water management, including rainwater harvesting and aquifer recharge (Medeazza, 2006; Srinivasan et al., 2010).

Desalination has been promoted as a 'resilience measure', particularly to drought. After initially being denied planning permission as an 'unsustainable' water supply, London's desalination plant opened in 2012 and is now characterised as providing resilience to drought. In Australia, the Gold Coast desalination plant, originally constructed but never used during the severe drought at the beginning of the twentieth century, has been represented as a 'resilience' measure after it was used during flood events when normal supplies were contaminated and water treatment works flooded. Although drought presents high uncertainty of rainfall, water shortages as a result of drought take months and years to evolve, providing time for construction and deployment of desalination when needed, instead of investing ahead of time in what becomes a 'stranded asset'. Drought resilience can be achieved by preparing contracts for construction that are ready to proceed should they be required, ensuring the complex planning and approvals are in place ahead of time (Turner et al., 2016). This can provide flexibility for rapid deployment of desalination without major capital investment in advance, which may prove unnecessary when normal rainfall returns. London, San Diego, Sydney, Melbourne and Adelaide have made considerable investment in desalination plants for drought resilience or during the midst of a severe drought event, which were never used. Mobile desalination plants are also promoted as providing greater flexibility for lower overall cost during periods of acute water scarcity, enabling resilience without the risk of investment in stranded assets (Ziolkowska, 2015).

Ecological modernisation

Affordable desalination, powered by renewable energy, with low environmental impact due to well-designed intake and brine discharge systems, is an archetypically modern solution to water shortages. The 2015 'Ecomodernist Manifesto' states:

> Urbanization, agricultural intensification, nuclear power, aquaculture, and desalination are all processes with a demonstrated potential to reduce human demands on the environment, allowing more room for non-human species.
>
> (Asafu-Adjaye et al., 2015, p. 18)

From an ecological modernisation viewpoint, the sustainability of desalination depends on its efficiency, environmental impacts and carbon emissions. The increase in efficiency and decrease in cost of desalination in recent decades makes this vision of an endless, sustainable supply of freshwater seem plausible in the near future.

Ecological modernisation draws attention to the potential for innovations in technologies and systems to solve environmental problems with minimal disruption to modern society. Desalination enables the continuation of existing water supply systems without requiring changes in consumption or lifestyles, as it simply provides a new source of water. As a capital-intensive, centralised supply it also fits within existing utility-driven infrastructure models, requiring minimal reform of regulations and institutions (Saurí and del Moral, 2001; Srinivasan et al., 2010).

Ecological modernisation promotes the role of the private sector in environmental innovation, using market-based instruments to allocate resources and create demand for new technologies and systems. The decreasing cost of desalination has been largely driven by private sector innovation, supported by university and other publically funded research and development. Economies of scale and the expansion of demand have further driven down costs of production of membranes. The private sector has also been involved in construction and operation of desalination, through public-private investment arrangements, contracts from public water utilities, and full private ownership. Desalination investment by private firms and in partnership with governments has driven expansion of the industry.

In Spain desalination was presented as a sustainable solution to ecological problems of over-abstraction of freshwater resource from the environment and preferable to the alternative of long-distance inter-basin transfers

(Saurí and del Moral, 2001; Swyngedouw, 2013). Similarly, in the US, Australia and China, desalination has been compared to water transfers, which have high pumping costs and energy requirements, and regional ecological impacts (Porter et al., 2015; Scarborough et al., 2015; Shahabi et al., 2017; Zheng et al., 2014; Ziolkowska and Reyes, 2016). In such situations desalination is a modern, advanced technology which moves beyond large, landscape-scale infrastructure projects that were prevalent throughout the nineteenth and twentieth centuries and were used in ancient civilisations. In Spain desalination was promoted by environmental campaigners and leftist politicians as a break with an ecologically destructive history of water management (Saurí and del Moral, 2001).

Similarly to potable reuse, desalination constructs freshwater as an industrial product, which can be manufactured by purifying a raw material. Freshwater is a limited resource, but limits can be overcome by technological advances which enable exploitation of water resources that were previously uneconomic to abstract. This is consistent with the conventional theory of economic substitution – as demand for a scarce resource increases and technology develops, previously uneconomic reserves become viable alternative supplies. The increased cost of 'manufactured' water is met by consumers or government subsidies and the direct beneficiaries of capital investment in new technologies are technology suppliers, engineering consultancies and utilities.

Desalination costs remain higher than conventional water supplies in most contexts. Desalination should then be considered as just one option for new water supply or drought resilience, to be compared against other options in terms of economic costs, environmental impacts, resource efficiency and social impacts. Decisions to install desalination plants ahead of lower-cost options or while conventional resources are still available undermine its credibility as a sustainable supply option (Turner et al., 2016). Desalination is attractive as a purely technical solution to water scarcity and may undermine the implementation of options which are lower cost and have lower environmental impacts and greater social benefits. For desalination to be considered sustainable from an ecologically modernist viewpoint, it must be considered as part of integrated urban water management, not as an isolated technical solution to water scarcity (Porter et al., 2015).

Socio-technical systems

As a supply-side option, abstracting water from the environment, desalination represents minimal change to the structure, function and experience of urban water systems. It provides a new source of water, and distribution, consumption and disposal can continue as before. Desalination re-enforces centralised models of urban water infrastructure, and its attractiveness to decision-makers faced with the uncertainty of drought is consistent with theories of socio-technical lock-in (Hommels, 2005). If cities are locked-in

to infrastructure pathways based on nineteenth-century models of centralised supply to meet growing demand, then desalination is the twenty-first-
century manifestation of the somewhat inevitable need to continuously
expand supply (March, 2015; Saurí and del Moral, 2001).

A socio-technical systems perspective also draws attention to the institutional structures that enable or constrain technologies. As a capital-intensive,
technical solution to urban water shortages, desalination is compatible with
existing models of engineering-led decision-making in urban water supply.
The increasing role of the private sector also supports the expansion of
desalination, as private capital can be raised for new investments and private sector technical expertise can provide direct, reliable solutions to water
resources uncertainty. International networks of experts, often working for
technology providers or engineering consultants, provide ready access to
knowledge about desalination and are able to rapidly deploy design and
construction teams as required by utilities and governments in need of
secure responses to water crises.

A broader analysis of desalination as a socio-technical system also draws
attention to the wider implications of its relatively high-energy demand.
Controversies about desalination in Australia, the UK, the US, Spain and
China show increasing public scrutiny of the water sector and concerns
about energy, costs and climate change (Dolnicar et al., 2011; Dolnicar
and Schäfer, 2009; Saurí and del Moral, 2001; Zheng et al., 2014). Socio-
technical analysis shows how controversies arise in science and technology,
demonstrating that desalination may not be the obvious solution to maintain modern patterns of water supply infrastructure that it seems. Debate
and controversy about desalination in politics and the media demonstrate
that water infrastructure is no longer perceived as a merely technical problem to be solved by engineers and utility managers. Energy, environment and
economic concerns from citizens, politicians, planners and civil society have
constrained the forecasts for rapid expansion of desalination.

Membrane technologies have been crucial in the expansion of RO, which
is now the predominant form of desalination. This represents a classic diffusion of innovation model of technical progress, as a technology that started
in university laboratories, was tested in small-scale pilot projects, benefited
from membrane technology and materials research in other sectors, and
reduced in cost due to growing economies of scale and reduced manufacturing costs.

Integration of desalination with power plants, through cogeneration and
the use of desalination to provide cooling water, demonstrates the complex interconnections in modern industrial systems. Desalination has not
emerged as an isolated technology to provide water for domestic use, but is
part of wider patterns of industrial development.

With more industrial sources of potable supply, water infrastructure technology becomes increasingly complex, requiring higher levels of technical
expertise to design and operate. Management of water systems remains in

the hands of large utilities, with a strengthening role for global engineering firms and technology suppliers. In highly risk-averse water infrastructure management, desalination is a very reliable way to avoid a deficit between supply and demand that could be harmful to public health.

Political ecology

The political ecology analysis of desalination focusses on the role of private capital in the expansion of the industry and its attractiveness as an option for water resource managers and politicians in maintaining existing socio-environmental hierarchies of power (March, 2015; Swyngedouw, 2015, 2013). Desalination remains more expensive and capital intensive than other responses to water shortages. From the point of view of political ecology the increasing role of private firms in water infrastructure provision, as owners, contractors, technology suppliers and partners, is a driving factor towards more capital-intensive solutions to water shortages and away from demand management or decentralised options (Bakker, 2010; Loftus and March, 2016).

Hug March and Alex Loftus point to the 'financialisation' of the water sector as underpinning the rationale for desalination (Loftus and March, 2016). Infrastructure investment is highly influenced by access to finance, and in turn financial institutions such as pension funds and sovereign wealth funds are attracted to infrastructure as a stable, long-term, secure investment. The availability of finance to invest in large-capital projects, and the profitability of contracts to build and operate desalination plants, is a significant driver in investment decision-making. Desalination is therefore a more attractive option to investors than less capital-intensive alternatives such as demand management. Globalisation of investment and trade and privatisation of water infrastructure have contributed to this trend. Multinational firms providing desalination technology and expertise benefit from government and privately owned water utilities looking for secure solutions to water shortages.

Political ecology analysis links the exploitation and pollution of the environment to capitalist systems of production that concentrate wealth and undermine equality. In this view, the expansion of desalination can be explained by its capacity to generate profit and investment returns to private capital. This has been facilitated by the privatisation of water infrastructure and the liberalisation of trade and investment, as much as it is a response to water shortages and technical developments. Teachers' pension funds from Canada, sovereign wealth from Qatar and an investment bank from Australia all enabled the financing of the desalination plant in London, with profit generated through customer bills within a regulated private monopoly for water supply (Loftus and March, 2016).

The financialisation of the water sector and its role in the expansion of desalination call into question its sustainability and the extent to which it represents a departure from the historic hydraulic paradigm which was underpinned by large-scale public sector investment in water infrastructure (March, 2015; Saurí and del Moral, 2001). The shift to private ownership, finance and contracts has intensified large-scale, centralised infrastructure solutions, enabling the private sector to benefit from the withdrawal of government from infrastructure services and constraints on the public sector to raise finance for investment. Desalination is therefore understood as a continuation of a dominant form of capital-intensive infrastructure provision, but with greater private profit than historic public utility models.

Radical ecology

From the viewpoint of radical ecology, desalination technologies perpetuate the view of nature as resource for human use, with both energy and water essentially limitless. The Promethean vision of turning salt water into fresh perpetuates a culture of consumption without limits and ignores wider environmental impacts of high water consumption, such as the energy required for wastewater treatment and the ecological impacts of wastewater discharges.

Radical ecological critique of modern industrial development and its domination and exploitation of nature presents a fundamental challenge to the enthusiasm for the expansion of desalination. Rather than seeking to understand natural hydrological and ecological processes, to live within limits and in partnership with nature, desalination perpetuates the modern, industrial approach to nature as a resource to be exploited for human use. Desalination brings the promise that humans can live on Earth without being limited by our basic need for water (Asafu-Adjaye et al., 2015). As such this is in direct contradiction to radical ecologist calls for restructuring society and technology to live in partnership with natural systems and as part of landscapes.

As radical ecologists seek to end the exploitation of nature and the damage to local environments, desalination also undermines connections to local environmental conditions. Adapting to a dry climate has been an important element of grassroots environmental ethics and sensibilities in post-colonial societies such as Australia and the US. Learning to live with the limits of local water resources has led to changes in gardening preferences towards local plants and changes in urban design (Syme et al., 2004). Desalination, in providing a limitless source of water irrespective of the local geography, runs counter to this sensibility. While the cost of desalination may still encourage water efficiency, the cultural message of an endless source of water provided by technology undermines nascent recognition of the interdependence between cities and their local hydrological and ecological contexts.

The prospect of a limitless supply of water from the sea as the solution to water crises overlooks the interrelationship with other natural and technical systems. Limitless water requires limitless energy, which is largely produced by fossil fuels. Increasing volumes of freshwater in supply also increases volumes of wastewater to be disposed of, with impacts on local hydrological systems from wastewater effluent and contaminants. The local environmental impacts of desalination itself are also significant from an ecological viewpoint, particularly concerns about the impact of intakes and brine discharge on local marine ecosystems.

Through energy consumption, wastewater production, and direct impacts, desalination demonstrates the fallacy of living beyond ecological limits. The idea of the ocean as a limitless source of freshwater comes at a cost of local environmental impacts and energy consumption. It also changes how cities and citizens relate to their local landscape and environment. Water has been a fundamental connection between citizens and their local landscapes, but with desalination water becomes a manufactured product produced by complex technology.

Desalination and sustainable cities

Is desalination sustainable? The short answer is no. Desalination is a sign that cities have exhausted conventional supplies and are living beyond hydrological limits. Even if it is powered by renewable energy, desalination perpetuates an imperative for control and exploitation of nature and the reinforcement of dominant modes of investment and management of urban water supply. By providing a seemingly straightforward technical solution to water scarcity, it draws attention away from broader movements towards integrated urban water management and efforts to reduce demand for water and energy.

Desalination is the ultimate ecological modernisation response to water insecurity, but it is not sustainable in any other framing, particularly those with a stronger critique of social and economic structures underpinning resource exploitation and environmental harm. Sustainability assessment may be a useful tool for evaluating desalination as a water resource option, and for identifying opportunities to improve the performance of particular plants. For instance, using renewable energy is preferable, that is, 'more sustainable', than powering desalination plants with fossil fuels. Desalination may also be less energy intensive or environmentally harmful than alternative supply options such as long-distance water transfers or exploitation of fossil groundwater resources. Decision support tools and quantitative evaluation methods that clarify trade-offs and comparison of options for water supply may be useful in planning water infrastructure, but they are unable to address more fundamental critiques of the sustainability of desalination as a technical fix to complex problems of water security in cities.

Desalination may yet have a role in sustainable cities as a resilience measure. Deployable, mobile desalination units may provide short-term supply when required during drought or other emergencies, without the high cost of expanding the total capacity of the supply system. Desalination plants that are constantly on standby, such as in London, Sydney and the Gold Coast, are costly to maintain and construct, while more decentralised, mobile desalination plants in the future may help avoid these costs while realising the benefits of rapid developments in desalination engineering in recent decades. Resilience planning methods should optimise regulatory, contractual and technical readiness to allow desalination to be constructed quickly when needed, reducing the risk of stranded assets that are installed and not used.

Desalination is not a sustainable solution to water scarcity if sustainability means reducing resource exploitation, living within hydrological and economical limits and ensuring affordable access to water for all. That doesn't mean desalination is always a bad idea or can't be justified for other reasons. Desalination provides a high certainty of supply of a resource that is critical to public health and economic activity in cities. Constant, reliable water supply is a fundamental requirement for modern urban living, and this objective may override sustainability goals in extreme water scarcity, at high economic and environmental cost.

References

Abutayeh, M., Li, C., Goswami, D.Y. and Stefanakos, E.K. 2014. Solar Desalination, in: Kucera, J. (Ed.), *Desalination*. John Wiley & Sons, Inc., Chichester, pp. 549–581.

Ahmed, M., Shayya, W.H., Hoey, D. and Al-Handaly, J. 2001. Brine disposal from reverse osmosis desalination plants in Oman and the United Arab Emirates. *Desalination* 133, 135–147. doi:10.1016/S0011–9164(01)80004–80007

Al-Gholaikha A., El-Ramly, N., Jamjooon, I. and Seaton, R. 1978. The World's First Large Seawater Reverse Osmosis Desalination Plant, at Jedda, Kingdom of Saudi Arabia. *Desalination* 27(3), 215–231.

Aly, N.H. and El-Figi, A.K. 2003. Mechanical Vapor Compression Desalination Systems – A Case Study. *Desalination, Desalination and the Environment: Fresh Water for All* 158, 143–150. doi:10.1016/S0011–9164(03)00444–00442

Arabindoo, P. 2011. Mobilising for Water: Hydro-Politics of Rainwater Harvesting in Chennai. *International Journal of Urban Sustainable Development* 3, 106–126. doi:10.1080/19463138.2011.582290

Asafu-Adjaye, J., Blomqvist, L., Brand, S., Brook, B., Defries, R., Ellis, E., Foreman, C., Keith, D., Lewis, M., Mark, L., Nordhaus, T., Pielke Jr, R., Pritzker, R., Roy, J., Sagoff, M., Shellenberger, M., Stone, R. and Teague, P. 2015. *An Ecomodernist Manifesto*. www.ecomodernism.org

Avrin, A-P., He, G. and Kammen, D.M. 2015. Assessing the Impacts of Nuclear Desalination and Geoengineering to Address China's Water Shortages. *Desalination* 360, 1–7. doi:10.1016/j.desal.2014.12.028

Bakker, K. 2010. *Privatizing Water*. Cornell University Press, Ithaca and London.

Belessiotis, V. and Delyannis, E. 2000. The History of Renewable Energies for Water Desalination. *Desalination* 128, 147–159. doi:10.1016/S0011–9164(00)00030–00038

Birkett, J. 2010. The History of Desalination Before Large-Scale Use – 01–003.pdf, in: *Encyclopedia of Desalination and Water Resources: History, Development and Management of Water Resources*. Encyclopedia of Life Support Systems (ELOSS) Publishers, Singapore, pp. 381–435.

Boals, C. 2009. Drinking from the Sea. *Circle of Blue*.

Burn, S., Hoang, M., Zarzo, D., Olewniak, F., Campos, E., Bolto, B. and Barron, O. 2015. Desalination Techniques – A Review of the Opportunities for Desalination in Agriculture. Desalination, *Desalination for Agriculture* 364, 2–16. doi:10.1016/j.desal.2015.01.041

Busch, M. and Mickols, W.E. 2004. Reducing Energy Consumption in Seawater Desalination. *Desalination* 165, 299–312. doi:10.1016/j.desal.2004.06.035

Cadotte, J.E. 1981. *Interfacially Synthesized Reverse Osmosis Membrane*. US4277344 A.

Cook, C. and Bakker, K. 2012. Water Security: Debating an Emerging Paradigm. *Global Environmental Change* 22, 94–102. doi:10.1016/j.gloenvcha.2011.10.011

Cooley, H. and Wilkinson, R. 2012. *Implications of Future Water Supply Sources for Energy Demands*. Water Reuse Association, Alexandria.

Delyannis, E. 2003. Historic Background of Desalination and Renewable Energies. *Solar Energy, Solar Desalination* 75, 357–366. doi:10.1016/j.solener.2003.08.002

Desalination and Water Reuse. 2015. Delayed Torrevieja Plant Pays Penalty. *Water Desalination and Reuse*. https://www.desalination.biz/news/0/Delayed-Torrevieja-plant-pays-penalty/7925/

Dobry, A. 1936. Les Perchlorates Comme Solvants de la Cellulose et de ses Derives. *Bulletin de La Societe Chimque de France* 5, 312–318.

Dolnicar, S., Hurlimann, A. and Grün, B. 2011. What Affects Public Acceptance of Recycled and Desalinated Water? *Water Research* 45, 933–943. doi:10.1016/j.watres.2010.09.030

Dolnicar, S. and Schäfer, A.I. 2009. Desalinated Versus Recycled Water: Public Perceptions and Profiles of the Accepters. *Journal of Environmental Management* 90, 888–900. doi:10.1016/j.jenvman.2008.02.003

El-Dessouky, H.T., Ettouney, H.M. and Al-Juwayhel, F. 2000. Multiple Effect Evaporation – Vapour Compression Desalination Processes. *Chemical Engineering Research and Design, Process Control* 78, 662–676. doi:10.1205/026387600527626

Elimelech, M. and Phillip, W. 2011. The Future of Seawater Desalination: Energy, Technology, and the Environment. *Science* 333, 712–717.

Ettouney, H., El-Dessouky, H. and Al-Roumi, Y. 1999. Analysis of Mechanical Vapour Compression Desalination Process. *International Journal of Energy Research* 23, 431–451. doi:10.1002/(SICI)1099–1114X(199904)23:5<431::AID-ER491>3.0.CO;2-S

Gebel, J. 2014. Thermal Desalination Processes, in: Kucera, J. (Ed.), *Desalination*. John Wiley & Sons, Inc., Chichester, pp. 39–154.

Ghaffour, N. 2009. The Challenge of Capacity-Building Strategies and Perspectives for Desalination for Sustainable Water Use in MENA. *Desalination and Water Treatment* 5, 48–53. doi:10.5004/dwt.2009.564

Ghaffour, N., Bundschuh, J., Mahmoudi, H. and Goosen, M.F.A. 2015. Renewable Energy-Driven Desalination Technologies: A Comprehensive Review on Challenges and Potential Applications of Integrated Systems. *Desalination, State-of-the-Art Reviews in Desalination* 356, 94–114. doi:10.1016/j.desal.2014.10.024

Ghaffour, N., Missimer, T.M. and Amy, G.L. 2013. Technical Review and Evaluation of the Economics of Water Desalination: Current and Future Challenges for Better Water Supply Sustainability. *Desalination* 309, 197–207. doi:10.1016/j.desal.2012.10.015

Gilron, J., Folkman, Y., Savliev, R., Waisman, M. and Kedem, O. 2003. WAIV – Wind Aided Intensified Evaporation for Reduction of Desalination Brine Volume. *Desalination, Desalination and the Environment: Fresh Water for All* 158, 205–214. doi:10.1016/S0011-9164(03)00453-00453

Glater, J. 1998. The Early History of Reverse Osmosis Membrane Development. *Desalination* 117, 297–309. doi:10.1016/S0011-9164(98)00122-00122

Global Clean Water Desalination Alliance. 2015. *G.C.W.D.A Joint Statement*, COP21, December 2015 [WWW Document]. Global Clean Water Desalination Alliance. http://frog4webservices.com/alliance2015/index.html.

Gude, V.G. 2016. Desalination and Sustainability – An Appraisal and Current Perspective. *Water Research* 89, 87–106. doi:10.1016/j.watres.2015.11.012

Hommels, A. 2005. Studying Obduracy in the City: Toward a Productive Fusion Between Technology Studies and Urban Studies. *Science, Technology & Human Values* 30, 323–351. doi:10.1177/0162243904271759

International Desalination Association. 2017. *Desalination by the Numbers*. International Desalination Association. http://idadesal.org/desalination-101/desalination-by-the-numbers/

Jacobsen, R. 2016. *Israel Proves the Desalination Era Is Here* [WWW Document]. Scientific American. www.scientificamerican.com/article/israel-proves-the-desalination-era-is-here/.

Kumar, M., Culp, T. and Shen, Y. 2017. Water Desalination: History, Advances and Challenges, in: *Frontiers of Engineering: Reports on Leading-Edge Engineering From the 2016 Symposium*. The National Academies Press, Washington, DC.

Lattemann, S. and Höpner, T. 2008. Environmental Impact and Impact Assessment of Seawater Desalination. *Desalination*, European Desalination Society and Center for Research and Technology Hellas (CERTH), Sani Resort 22–25 April 2007, Halkidiki, Greece 220, 1–15. doi:10.1016/j.desal.2007.03.009

Lee, H.-J. and Moon, S-H. 2014. Electrodialysis Desalination, in: Kucera, J. (Ed.), *Desalination*. John Wiley & Sons, Inc., Chichester, pp. 287–326.

Lind, J. 1811. *An Essay on Diseases Incidental to Europeans in Hot Climates with the Method of Preventing Their Fatal Consequences*, First American, from the sixth London edition. ed. William Duane, Philadelphia, PA.

Loeb, S. 1981. The Loeb-Sourirajan Membrane: How It Came About. *ACS Symposium Series* 153(1), 1–9.

Loeb, S. and Sourirajan, S. 1963. Sea Water Demineralization by Means of an Osmotic Membrane. *Advances in Chemistry* 38(9), 117–132.

Loftus, A. and March, H. 2016. Financializing Desalination: Rethinking the Returns of Big Infrastructure. *International Journal of Urban and Regional Research* 40, 46–61. doi:10.1111/1468-2427.12342

MacGowan, C. 1963. History, Function and Programme of the Office of Saline Water, in: *Water Resources Review*. United States Geological Survey, Washington, D.C., pp. 24–32. https://nmwrri.nmsu.edu/wp-content/uploads/2015/watcon/proc8/MacGowan.pdf

Mackey, E., Pozos, N., James, W., Seacord, T., Hunt, H. and Mayer, D. 2011. *Assessing Seawater Intake Systems for Desalination Plants*. Water Research Foundation, Denver.

March, H. 2015. The Politics, Geography, and Economics of Desalination: A Critical Review. *WIREs Water* 2, 231–243. doi:10.1002/wat2.1073

Medeazza, G.M. von. 2006. Desalination in Chennai: What About the Poor and the Environment? *Economic and Political Weekly* 41, 949–952.

Megahed, M.M. 2001. Nuclear Desalination: History and Prospects. *Desalination* 135, 169–185. doi:10.1016/S0011-9164(01)00148-00145

Megahed, M.M. 2003. An Overview of Nuclear Desalination: History and Challenges. *International Journal of Nuclear Desalination* 1, 2–18.

Miri, R. and Chouikhi, A. 2005. Ecotoxicological Marine Impacts from Seawater Desalination Plants. *Desalination, Desalination and the Environment* 182, 403–410. doi:10.1016/j.desal.2005.02.034

Porter, M.G., Downie, D., Scarborough, H., Sahin, O. and Stewart, R.A. 2015. Drought and Desalination: Melbourne Water Supply and Development Choices in the Twenty-First Century. *Desalination and Water Treatment* 55, 2278–2295. doi:10.1080/19443994.2014.959743

Reid, C.E. and Breton, E.J. 1959. Water and Ion Flow Across Cellulosic Membranes. *Journal of Applied Polymer Science* 1, 133–143. doi:10.1002/app.1959.070010202

Rothausen, S.G.S.A. and Conway, D. 2011. Greenhouse-Gas Emissions from Energy Use in the Water Sector. *Nature Climate Change* 1, 210–219. doi:10.1038/nclimate1147

Runte, G. 2016. *Seawater and Brackish Water Desalination*. BCC Research. https://www.bccresearch.com/

Saurí, D. and del Moral, L. 2001. Recent Developments in Spanish Water Policy: Alternatives and Conflicts at the End of the Hydraulic Age. *Geoforum* 32, 351–362. doi:10.1016/S0016-7185(00)00048-00048

Scarborough, H., Sahin, O., Porter, M. and Stewart, R. 2015. Long-Term Water Supply Planning in an Australian Coastal City: Dams or Desalination? *Desalination* 358, 61–68. doi:10.1016/j.desal.2014.12.013

Shahabi, M.P., McHugh, A., Anda, M. and Ho, G. 2017. A Framework for Planning Sustainable Seawater Desalination Water Supply. *Science of the Total Environment* 575, 826–835. doi:10.1016/j.scitotenv.2016.09.136

Shiklomanov, I. 2000. Appraisal and Assessment of World Water Resources. *Water International* 25(1), 11–32.

Srinivasan, V., Gorelick, S.M. and Goulder, L. 2010. Sustainable Urban Water Supply in South India: Desalination, Efficiency Improvement, or Rainwater Harvesting? *Water Resource Research* 46, W10504. doi:10.1029/2009WR008698

Subramani, A. and Jacangelo, J.G. 2015. Emerging Desalination Technologies for Water Treatment: A Critical Review. *Water Research* 75, 164–187. doi:10.1016/j.watres.2015.02.032

Surwade, S.P., Smirnov, S.N., Vlassiouk, I.V., Unocic, R.R., Veith, G.M., Dai, S. and Mahurin, S.M. 2015. Water Desalination Using Nanoporous Single-Layer Graphene. *Nature Nanotechnology* 10, 459–464. doi:10.1038/nnano.2015.37

Swyngedouw, E. 2013. Into the Sea: Desalination as Hydro-Social Fix in Spain. *Annals of the Association of American Geographers* 103, 261–270. doi:10.1080/00045608.2013.754688

Swyngedouw, E. 2015. *Liquid Power: Contested Hydro-Modernities in Twentieth-Century Spain*. MIT Press, Cambridge, MA.

Syme, G.J., Shao, Q., Po, M. and Campbell, E. 2004. Predicting and Understanding Home Garden Water Use. *Landscape and Urban Planning* 68, 121–128. doi:10.1016/j.landurbplan.2003.08.002

Tang, C.Y., Zhao, Y., Wang, R., Hélix-Nielsen, C. and Fane, A.G. 2013. Desalination by Biomimetic Aquaporin Membranes: Review of Status and Prospects. *Desalination, New Directions in Desalination* 308, 34–40. doi:10.1016/j.desal.2012.07.007

Turner, A., White, S., Chong, J., Dickinson, M., Cooley, H. and Donnelley, K. 2016. *Managing Drought: Learning from Australia.* Pacific Institute, Oakland.

Vedachalam, S. 2012. *Water Supply in Chennai: Desalination and Missed Opportunities* (SSRN Scholarly Paper No. ID 2057301). Social Science Research Network, Rochester, NY.

Virgili, F. and Pankratz, T. 2016. *IDA Desalination Yearbook 2016–2017.* International Desalination Association, London.

Voutchkov, N. 2016. *Desalination – Past, Present and Future.* International Water Association, London.

Water Corporation. 2017. *Southern Seawater Desalination Plant* [WWW Document]. Water Corporation of WA. www.watercorporation.com.au/water-supply/our-water-sources/desalination/southern-seawater-desalination-plant.

Wilf, M. 1997. Design Consequences of Recent Improvements in Membrane Performance. *Desalination* 113, 157–163. doi:10.1016/S0011–9164(97)00124–00120

Wilf, M. 2014. The Reverse Osmosis Process, in: Kucera, J. (Ed.), *Desalination.* John Wiley & Sons, Inc., Chichester, pp. 155–204.

Zeitoun, M., Lankford, B., Krueger, T., Forsyth, T., Carter, R., Hoekstra, A.Y., Taylor, R., Varis, O., Cleaver, F., Boelens, R., Swatuk, L., Tickner, D., Scott, C.A., Mirumachi, N. and Matthews, N. 2016. Reductionist and Integrative Research Approaches to Complex Water Security Policy Challenges. *Global Environmental Change* 39, 143–154. doi:10.1016/j.gloenvcha.2016.04.010

Zheng, X., Chen, D., Wang, Q. and Zhang, Z. 2014. Seawater Desalination in China: Retrospect and Prospect. *Chemical Engineering Journal* 242, 404–413. doi:10.1016/j.cej.2013.12.104

Ziolkowska, J.R. 2015. Is Desalination Affordable? – Regional Cost and Price Analysis. *Water Resource Management* 29, 1385–1397. doi:10.1007/s11269–11014–10901-y

Ziolkowska, J.R. 2016. Desalination Leaders in the Global Market – Current Trends and Future Perspectives. *Water Science and Technology: Water Supply* 16, 563–578. doi:10.2166/ws.2015.184

Ziolkowska, J.R. and Reyes, R. 2016. Impact of Socio-Economic Growth on Desalination in the US. *Journal of Environmental Management* 167, 15–22. doi:10.1016/j.jenvman.2015.11.013

10 Conclusion

Water and sustainable cities

There is no universal roadmap for urban water sustainability. Water infrastructures reflect political, social and economic priorities, coming together in complex and conflicting ways within debates, designs and plans for sustainability. Within any given city, there will be competing visions for sustainability and water, and different strategies for implementing change.

Urban water sustainability reflects wider discourse and the complex interplay between technologies and values. Sustainability is a social, political and technical choice. Choices about technologies and infrastructure are made within social, political and cultural frameworks, which are both constraining and dynamic. There are choices to be made in how to frame sustainability according to different values and knowledge systems, and choices between different technical options. Choices have consequences – for water, people, cities and nature. There are always alternatives. Sustainability is not inevitable.

This book presented five frameworks – sustainable development, ecological modernisation, socio-technical systems, political ecology and radical ecology – for assessing five specific technical trends – demand, sanitation, drainage, reuse and desalination – in urban water sustainability. The purpose has been to make the political and technical options available to cities more explicit, rather than outline a blueprint for a new paradigm of water management. In conclusion, this chapter reviews each of the technical trends that have emerged in urban water sustainability research and practice in recent decades. It assesses the usefulness of analysing technologies through different frameworks and the implications for decisions about urban infrastructure. It finally concludes with reflections on how better understanding society, politics and technology might help deliver sustainable cities that work for people, water and nature.

Technologies

Looking at sustainability through technologies might seem to run counter to the ideal of integration across sectors and disciplines and the need to emphasise social, economic and political factors. However, if sustainability and infrastructure are outcomes of relationships between these different factors, then it becomes difficult to know where to begin. Analysing significant technical trends in urban water management provided a starting point from which to explore the complexity of urban water infrastructure and sustainability.

Prominent debates and developments in urban water technologies reveal significant common ground across different conceptual framings of sustainability. It is universally agreed that everyone deserves access to water and sanitation. It is now recognised as a human right by the United Nations and is prominent in the Sustainable Development Goals (United Nations, 2015; United Nations General Assembly, 2010). Water is essential for good health, and decent toilets are a prerequisite for human dignity (Greed, 1996; UN-HABITAT, 2003). It is also uncontroversial to say that water should be used wisely (Butler and Memon, 2005). Reducing water wastage is a basic responsibility for utilities and citizens. Protecting homes and streets from surface water flooding and reducing pollution from runoff is a basic infrastructure service (Butler and Davies, 2010). Green infrastructure presents opportunities, but also uncertainties, to deliver multiple benefits in sustainable cities (Dover, 2015). Reusing and recycling water, like any resource, is likely to be part of sustainable cities, though lifecycle impacts and public acceptability must be included in evaluation of technical options at different scales (Bixio et al., 2008). Desalination may be necessary in some cities and should be designed to minimise environmental and climate impacts, but it is not sustainable (March, 2015).

The same technology can be used to support different concepts about how society is organised and the meaning of water in cities. Rainwater harvesting may be a tool for integrated, shared management of surface water and water resources, or it can be a means for individual households to secure their own source of water for private use in times of scarcity (Campisano et al., 2017). Reuse can be judicious use of water resources within local hydrological limits supporting a new environmental ethic, or a technical option for new water supply that represents minimal disruption to existing infrastructure and consumption (Bell and Aitken, 2008).

Technologies can also take on very different meanings depending on how they are framed by political and social relations. Container-based sanitation may be an innovative technology and business model within ecological modernisation, but it has been representative of gross social inequalities as experienced by poor and marginalised communities and analysed through the lens of political ecology (McFarlane and Silver, 2017; Sanergy, 2017). An ecological modernist may view water metering as a rational, objective source of information to inform behaviour change, while a socio-technical systems analysis shows that data about volumes of water flowing into a house bears very little relation to how people experience water in their everyday lives (Boyle et al., 2013; Shove, 2010).

The physical form and performance of different technologies are important in achieving sustainability. However, technology choice is not merely a matter of assessing and optimising physical parameters. Technologies and infrastructures mediate between people and the environment, and the form of technology reveals the state of relations between cities and nature.

Frameworks

Sustainability is notoriously difficult to define. Enough is shared between different researchers, policy-makers and practitioners in the field to identify collectively as 'urban water sustainability', but within the overall movement there are sufficient differences to identify five alternative frameworks – sustainable development, ecological modernisation, socio-technical systems, and political ecology and radical ecology. There is some overlap between different frameworks, but also clear distinctions and points of incompatibility. Within the five frameworks, sustainability is variously:

- Human development within ecological limits (sustainable development)
- Rational problem of efficiency, optimisation and resource allocation (ecological modernisation)
- Transition and co-evolution of material relationships between people, technologies and the environment (socio-technical systems)
- Redistribution of resources and power to achieve fair socio-ecologies and urban metabolisms (political ecology)
- Repairing human relationships with nature and each other (radical ecology)

The frameworks provide alternative pathways and strategies for alleviating agreed problems. For instance, poverty and inequality are accepted as issues for sustainability to address. In sustainable development they are solved through economic growth, within environmental constraints. Similarly, for ecological modernisation poverty is best addressed through continued innovation and industrialisation, improving overall benefits from development. For socio-technical systems perspectives poverty might be considered as a problem of equity in accessibility to infrastructural systems. Political ecology points out the underlying political and economic structures and trends that drive unequal outcomes, particularly the forces of global capitalism and neoliberal policy doctrine. For radical ecologists poverty may best be alleviated through fundamentally reorganising society away from industrialisation towards communities that are embedded in local landscapes and ecosystems.

Similarly, it is widely agreed that institutional as well as technical change is required to achieve sustainability. For sustainable development strengthening national and international institutions of governance and finance is needed to deliver agreed goals. Ecological modernisation promotes reform of existing institutions, including the market, the state, and science and technology, to incorporate environmental values. Socio-technical systems perspectives highlight the need for institutions to co-evolve with changing technical form of infrastructure, and that innovation is needed in institutions and cultures as well as technologies. Restructuring the global economy, and national and local democracy, to redistribute power is needed to ensure socio-ecological equity, according to political ecology. Radical ecology calls for more fundamental reorganisation of society and politics, based upon local ecological communities.

Each of the frameworks presents different priorities and identifies different gaps in knowledge about sustainability. The frameworks reflect different values, but they can also highlight blind spots or unintended consequences of widely held beliefs. Different frameworks ask different questions of the empirical data. For instance, ecological modernists concerned with efficiency may be interested in the relationship between consumption and water pricing, while political ecologists will draw attention to the different impacts that water pricing has on those on low or high incomes. One framework might provide an analysis of the aggregate costs and benefits of water metering and pricing, while the other draws out the uneven distribution of those costs and benefits across society and the environment.

Technical research into specific technologies and engineering models is often framed within a particular set of values and assumptions about the role of technology in society, but it is rarely acknowledged. In some instances, researchers make a strategic, explicit choice to position research within a particular framework to maximise policy relevance. In others it might be an unconscious response to disciplinary norms and the requirements of research funders and partners. As researchers are driven to demonstrate the impact of their work through partnership with industry and government, they may align their work with a dominant framework. For others, changing the discourse itself or constructing new ways of understanding urban water sustainability is perhaps the highest possible impact, shaping and defining new paradigms.

Constructing infrastructure

There is no technically optimal solution and no clear political consensus about the form of urban water sustainability. There is constant interaction as technology and politics shape each other, making some things seem inevitable and other things unthinkable. In debates about infrastructure, the physical possibilities of what technology can deliver, the material constraints of what natural systems can sustain and the social values of how we want to live are all important. Some of these things can be known using engineering science and methods that are quantifiable. Some must be debated and negotiated. Others require both numbers and negotiation, particularly discussions about risk, equity, service levels, wellbeing and environmental health.

Deliberation about urban water sustainability, as in many things, can become polarised as each participant claims to be 'rational', 'objective' and 'reasonable', while claiming that others are 'political', 'subjective' and 'out of touch'. Proponents of sustainable systems have claimed that dominant approaches to infrastructure are locked-in to risk-averse cultures that perpetuate outdated beliefs about technology and the environment (Brown and Farrelly, 2009). Conversely, they are accused of pushing a particular version of environmental politics onto professions and societies that remain primarily concerned with economic growth, public safety and infrastructure reliability (Marlow et al., 2013). Both are right.

In deciding upon options to pursue for sustainable urban water infrastructure the choice is not between an objective technical rationality and a political ideology. The choice is between different assemblages of technologies and values, which are constantly emerging and evolving. Different framings lead to different analysis of technologies. What may appear to be a rational inevitability from within one framework might be seen as a political programme from another. For instance, water metering and charging are entirely rational within an ecological modernist framework, while political ecologists claim that these strategies enforce a neoliberal ideology which leads to increasing social and environmental inequality. From a sociotechnical systems point of view, sustainable drainage is a niche technical innovation requiring institutional reform to become standard design practice across cities, whilst an ecological modernist framing might see it as an inefficient and expensive expression of radical environmental values.

Politics, values and conceptual frameworks are always present in decisions about infrastructure in cities. High-quality deliberation and debate about the future of cities requires that both politics and technology can be discussed at the same time. Engineers, planners and designers therefore need to be conscious of and skilled in negotiating discourse as well as technology. Revealing the conceptual framing of particular propositions for change enables clearer assessment of consequences for cities and nature.

Urban water sustainability

Water is a fundamental element of built and natural environments, supporting life, livelihoods, culture, wellbeing and development. Water is simultaneously a consumptive resource, an infrastructure service, the foundation of ecosystems, an aesthetic element in design, a source of personal, cultural and spiritual comfort and contemplation, and essential for good public health. It connects cities to landscapes and local environments. Water is heavy, easily polluted and energy intensive to purify, making it difficult to transport and complex to treat. While food, fibre, energy and other materials are relatively easy to import into cities from far away, water mostly still comes from local catchments. Water plays a fundamental role in place-making, through urban design, gardening and lifestyles that adapt to wet and dry seasons. It is also part of our most intimate, private habits and everyday lives, lived out in bathrooms, kitchens and laundries across the city. It is a natural resource, an essential service and a cultural symbol.

Water has a unique role within sustainable cities. There are obvious parallels between managing water and managing other urban infrastructures such as energy, transport and waste, including reducing demand, improving efficiency, ensuring fair access, decentralising supply, and remediating pollution and degradation. But water is more than this. It is a unique integrating element which points to the potential not only to improve urban efficiency but to fundamentally reconsider how cities relate to nature. Through water, ideas of urban sustainability as the means of reducing human impacts, restoring ecosystems, supporting wellbeing and learning to live within local and global limits are made real and immediate. These connections are not merely abstract, global, technical or economic relations; they are tangible, visceral, visible and local.

How to achieve urban water sustainability is the subject of technical and political debate, alongside practical implementation and experimentation. It is tempting to try to find a unifying framework to integrate divergent strategies for delivering infrastructure to manage water in cities to provide for human health and wellbeing within hydrological and ecological limits. This temptation should be resisted. Diversity brings resilience. Alternate framings of sustainability provide checks and balances and reveal gaps in knowledge and justice. If infrastructures are social and political as well as technical systems, then it is important to explore the full range of options and their consequences when deciding upon the future of water in cities. Over time, new options appear while others fall away, but the capacity to recognise and discuss the technical, political, ecological and social implications of particular choices provides the foundation for more transparent and robust decision-making about this most vital element of cities and nature.

References

Bell, S. and Aitken, V. 2008. The Socio-Technology of Indirect Potable Water Reuse. *Water Science & Technology: Water Supply* 8, 441. doi:10.2166/ws.2008.104

Bixio, D., Thoeye, C., Wintgens, T., Ravazzini, A., Miska, V., Muston, M., Chikurel, H., Aharoni, A., Joksimovic, D. and Melin, T. 2008. Water Reclamation and Reuse: Implementation and Management Issues. *Desalination* 218, 13–23. doi:10.1016/j.desal.2006.10.039

Boyle, T., Giurco, D., Mukheibir, P., Liu, A., Moy, C., White, S. and Stewart, R. 2013. Intelligent Metering for Urban Water: A Review. *Water* 5, 1052–1081. doi:10.3390/w5031052

Brown, R. and Farrelly M. 2009. Delivering Sustainable Urban Water Management: A Review of the Hurdles We Face. *Water Science and Technology* 59(5), 839–846.

Butler, D. and Davies, J. 2010. *Urban Drainage*, Third Edition. CRC Press, London.

Butler, D. and Memon, F.A. (Eds.) 2005. *Water Demand Management*. IWA Publishing, London.

Campisano, A., Butler, D., Ward, S., Burns, M.J., Friedler, E., DeBusk, K., Fisher-Jeffes, L.N., Ghisi, E., Rahman, A., Furumai, H. and Han, M. 2017. Urban Rainwater Harvesting Systems: Research, Implementation and Future Perspectives. *Water Research* 115, 195–209. doi:10.1016/j.watres.2017.02.056

Dover, J.W. 2015. *Green Infrastructure: Incorporating Plants and Enhancing Biodiversity in Buildings and Urban Environments*. Routledge, London and New York.

Greed, C.H. 1996. Planning for Women and Other Disenabled Groups, with Reference to the Provision of Public Toilets in Britain. *Environment and Planning A* 28, 573–588. doi:10.1068/a280573

March, H. 2015. The Politics, Geography, and Economics of Desalination: A Critical Review. *WIREs Water* 2, 231–243. doi:10.1002/wat2.1073

Marlow, D., Moglia, M., Cook, S. and Beale, D. 2013. Towards Sustainable Urban Water Management: A Critical Reassessment. *Water Research* 47(20), 7150–7161.

McFarlane, C. and Silver, J. 2017. The Poolitical City: "Seeing Sanitation" and Making the Urban Political in Cape Town. *Antipode* 49, 125–148. doi:10.1111/anti.12264

Sanergy. 2017. *Sanergy* [WWW Document]. http://saner.gy/.

Shove, E. 2010. Beyond the ABC: Climate Change Policy and Theories of Social Change. *Environment and Planning A* 42, 1273–1285. doi:10.1068/a42282

UN-HABITAT. 2003. *Water and Sanitation in the World's Cities*. Earthscan, London and Sterling, VA.

United Nations. 2015. *Sustainable Development Goals* [WWW Document]. Sustainable Development Knowledge Platform, Department of Economic and Social Affairs. https://sustainabledevelopment.un.org/?menu=1300.

United Nations General Assembly. 2010. *The Human Right to Water and Sanitation* (Resolution adopted by the General Assembly No. A/RES/64/292). United Nations, New York.

Index

actor-network theory (ANT) 42–43
agriculture 15, 17, 139, 161
air pollution 40
Ajzen's Theory of Planned Behaviour 71
Aktau, Kazakhstan 160
alternative technology movement 29
anthropocentrism 48
appliances 66–68
appropriate technology movement 29

behaviour change 70–71, 75
Behaviour Change Wheel 71
blue infrastructure 116
Bombay/Mumbai, India 100
Breakthrough Institute 41
building codes 66, 75, 123, 141

Cape Town, South Africa 100
capitalism 40–41
catchment-based integrated
    management 17
catchments 1–2, 40–41
Chennai, India 160, 162
China 121–122
climate change 15, 111
combined sewer overflows (CSOs)
    90–91, 106–107, 112, 120, 123
Community-Led Total Sanitation
    (CLTS) movement 92
composting toilets 94
critical ecofeminism 49
cultures 72–73

decentralisation 28–31, 144
deep ecology 48–49, 126
demand: behaviour change 70–71, 75;
    building codes 66, 75; components
    of 59–60; cultures 72–73; ecological
    modernisation 75; end-use studies 62;

factors shaping 60–62; fittings and
appliances 66–68; forecasting 59–63;
frameworks 73–79, 174; impact
of climate change 15; labelling 70;
leakage 64–66, 74; management
6, 56–58, 63–73, 76, 79; metering
68–69, 74, 77; New York City water
demand management 67; political
ecology 77–78; practices 71–72,
76; radical ecology 78–79; reuse/
recycling and 132; socio-technical
systems 76–77; strategies for
reducing 57; in sustainable cities 79;
tariffs 69–70, 74, 77; theories of
64; water scarcity and 15; water use
patterns 57, 59–60, 77; water use
restrictions 64
desalination: controversies about
    165; ecological modernisation
    163–164; energy systems
    159–161; environmental impacts
    159–161; financialisation of 163,
    165–167; frameworks 161–168,
    174; integrated water resources
    management 17; membrane
    157–159, 165; methods 154–161;
    nuclear 160–161; overview of
    152–154; political ecology 166–167;
    potable reuse and 135; privatisation
    167; radical ecology 167–168;
    socio-technical systems 164–166;
    sustainable cities and 168–169;
    sustainable development 8, 161–163;
    technologies 6, 154–161, 163–164,
    175; thermal 154–157
digital infrastructure 30–31
direct potable water reuse 134–135
disaster preparedness, 15
discourse analysis 35–36

disease 85–87
dishwashers 67
drainage: climate change 111; core
    element of urban upgrading
    programmes, 118–119; ecological
    modernisation 119–124; frameworks
    117–126, 174; infrastructure 7,
    25, 106–127; management 114;
    political discourse 125; political
    ecology 124–125; radical ecology
    125–126; rainfall 107–112; runoff
    106, 107–112; sustainable 106–107,
    112–117, 125–126; for sustainable
    cities 107, 126–127; sustainable
    development 118–119; technologies
    6, 112; water quality 111–112
drought 26, 79, 134, 140, 162
dual-reticulation reuse systems
    136–138, 140–141

earth closet 87, 91
ecological feminism 48–50
ecological modernisation: definition of
    5; demand 75; desalination 163–164;
    drainage 119–124; overview of
    39–41; reuse/recycling 140–141;
    sanitation 97–98; socio-technical
    systems 141–143; use of framework
    176–177
'Ecomodernist Manifesto' 41
ecosystem services 119–120
electrodialysis 157
energy management 63
energy systems 17–18, 25, 133, 138,
    141, 159–161, 165, 168
environmental movement 37, 40, 48
essentialist ecofeminism 49

filtration 136
'fit-for-purpose' water systems 131, 138
fittings 66–68
flocculation 136
frameworks: conceptual 35–36;
    demand 73–79; desalination
    161–168; drainage 117–126;
    ecological modernisation 5, 39–41,
    75, 97–98, 119–124, 140–141,
    163–164, 174, 176; explanatory
    mode 50; normative mode 50;
    political ecology 6, 44–48, 77–78,
    99–101, 124–125, 143–144,
    166–167, 174, 176; pragmatic
    mode 50–52; radical ecology 6,

48–50, 101, 125–126, 144–145,
    167–168, 174, 176; reuse/recycling
    139–145; sanitation 95–101; socio-
    technical systems 5–6, 41–44,
    76–77, 98–99, 164–166, 174, 176;
    sustainable development 5, 36–39,
    73–75, 95–97, 118–119, 139–140,
    161–163, 174, 176; types of 5–6,
    35–52; of urban water sustainability
    50–52; use of 6
'Fresh Life Toilets' 94

gender factors 31–32
Global North: drinking water
    contamination 27; integrated water
    resources management 18; issue of
    resource management and pollution
    control 12; lack of investment in
    infrastructure 28; managing water
    demand 57–58, 60, 74, 79; principles
    of appropriate technology 29;
    sustainable development 37–38;
    sustainable drainage 119, 123; water-
    based sanitation systems 7, 84–85,
    87, 90–94, 97–98, 99, 100, 102
Global South: absence of infrastructure
    25–26; integrated water resources
    management 17; managing water
    demand 3, 6, 56–57, 75, 79;
    principles of appropriate technology
    29; reuse of water 139; role of
    private sector in water infrastructure
    provision 47; sanitation
    infrastructure, 100, 102; sustainable
    development 37–38, 161; urban
    drainage 125; water-based sanitation
    systems 85, 90, 102; waterless
    sanitation 92–94
Global Water Partnership 16
Gold Coast, Australia 137, 162, 169
grassroots innovation 29
'Green City, Clean Waters' programme
    120–121, 123
green infrastructure 19, 66, 116, 175
green roofs 114
greywater 133, 137–138, 144

Honey Bee Networks 29

illegal use 59
India 29
indirect potable water reuse 135
industrialisation 37, 85

industrial wastewater 133
infrastructure: blue 116; case for
    unsustainability of existing 2–3;
    challenges of governance and
    financing 122; constructing 4–5,
    24–33, 177–178; consumption
    practices and 76; conventional
    19–20, 24–25; decentralisation
    28–31; definition of 43; desalination
    161–168; digital 30–31; distributed
    technologies 116–117; drainage 7,
    25, 106–127; financialisation of
    28, 77, 165–167; frameworks 5–6,
    35–52; gender factors 31–32; green
    15, 19–20, 66, 116, 175; integration
    polices 16–18; linear pattern of
    131; obduracy of 43; ownership
    27–28; planning 59; privatisation
    167; privatisation of 27–28,
    46–48, 77; recent trends towards
    fragmentation of 100–101; reuse/
    recycling 132, 139–141; sanitation
    85–94, 100; socio-technical systems
    41–44; sustainability of 1–2, 6–7;
    sustainable cities and 32–33; urban
    background 25–27; wastewater
    87–91, 133; water reuse 141–143,
    146; water sensitive cities 18–20
integrated urban water management
    (IUWM) 1, 17–18
integrated water resources management
    (IWRM) 3, 16–17, 31, 136–137
International Conference on Water and
    the Environment 13
international water deliberation 13–16

labelling 70
large technical systems (LTS), 42
Latin America 29
latrines 92–93, 98–99
leakage 59, 64–66, 74
'light green' technologies 145
London, England 112, 160, 169
Loowatt system 94, 96
low impact development (LID) 107,
    112

market-based economics 40–41
membrane bioreactors (MBR) 136–137
membrane desalination 157–159, 165
metering 68–69, 74, 77, 175
multi-effect distillation (MED)
    154–156, 160

multi-stage flash (MSF) distillation
    154–156, 160
municipal wastewater 85, 92, 131–134,
    136, 138

neoliberal policies 27, 40
New York City 66, 67
non-dual-reticulation reuse systems 137
non-potable reuse 132–133, 135–138,
    141, 145
non-revenue water 59
nuclear desalination 160–161

obduracy 43
operational water 59
ownership 27–28

pail system 91
permeable pavements 114–115
pesticides 37
Philadelphia, Pennsylvania 120–121,
    123–124
Pimpama Coomera, Australia 137
policy 3, 16–18
political ecology: definition of 6;
    demand 77–78; desalination
    166–167; drainage 124–125;
    overview of 44–48; reuse/recycling
    143–144; sanitation 99–101; use of
    framework 176–177
pollution 14, 37, 40, 84, 107, 111, 113,
    166, 175
potable reuse 132–135, 140–143
practices 71–72, 76
practice theories 42
privatisation 27–28, 46–48, 77, 163,
    165

radical ecology: definition of 6; demand
    78–79; desalination 167–168;
    drainage 125–126; overview of
    48–50; reuse/recycling 144–145;
    sanitation 101; use of framework
    176–177
rainfall 107–112
rain gardens 115
rainwater 19, 114, 133, 137–138, 139,
    144, 175
regulation 98, 112
renewable energy 160
resilience 140, 162–163, 168–169
reuse/recycling: challenges for 145–146;
    ecological modernisation 140–141;

environmental impacts 138–139; forms 132–138; frameworks 139–145, 174; infrastructure 132; integrated water resources management 17; non-potable 132–133, 135–138, 141, 145; political ecology 143–144; potable 132, 134–135, 140–143; public concerns 141–142; sustainable cities and 145–146; sustainable development 7–8, 139–140; technologies 6, 132, 137–138, 140–145, 175; wastewater 7
reverse osmosis (RO) treatment 134, 154, 157–159, 160, 161, 165
runoff 106, 107–112, 175

sanitation: challenges for water-based 89–91; challenges of providing 3–4; development of water-based 85–88; dry 91, 101; ecological 91–92, 94, 97–98, 101; ecological modernisation 97–98; frameworks 95–101, 174; infrastructure 85–94, 100; Millennium Development Goals for 14–15, 84; political ecology 99–101; politics of 100–101; radical ecology 101; regulation for water treatment discharge 98; sewage treatment methods 88–89, 98–99; socio-technical systems 98–99; in sustainable cities 102; sustainable development 95–97; Sustainable Development Goals 95–96; technical alternatives 102; technologies 6–7, 175; water-based 7, 85–91, 96, 98–99, 101, 102; waterless 91–94
sewage sludge 90–91
sewers 7, 87, 89–91, 119
Slum Networking approach 119
social construction of technology (SCOT) 42
social ecology 48–49
socio-technical systems: definition of 5–6; demand 76–77; desalination 164–166; ecological modernisation 141–143; overview of 41–44; sanitation 98–99; use of framework 176–177
Sponge City programme 121–122
stormwater 19, 106, 107–112, 133
stormwater charging 121
surface water drains 7

sustainable development: definition of 5; demand 73–75; desalination 161–163; drainage 118–119; overview of 36–39; reuse/recycling 139–140; sanitation 95–97; use of framework 176–177
sustainable drainage systems (SuDS) 107, 112, 126
sustainable urban water management (SUWM): aim of 1; frameworks 5–6; technologies 6–8
swales 115–116
Sydney, Australia 138, 160, 169

tariffs 69–70, 74, 77
technologies: alternative 29–30; appropriate 29–30; choice of 174–176; decentralisation of 29–30; demand management 77; desalination 154–161, 163–164; distributed 116–117; drainage 112–117; reuse/recycling 132, 137–138, 140–145; role in ecological modernisation 39–40; role in sustainability 6–8; sanitation 85–94, 102; socio-technical systems 42–44
Thames Barrier 110
thermal desalination 154–157
toilets 26, 66–67, 76, 85, 89, 93–94, 97–99, 175
Toowoomba, Australia 135
transitions theory 44, 124

unbilled use 59
United Nations (UN): Agenda 21 13–14, 38, 118; Commission for Environment and Development Report 13; Commission on Sustainable Development 14; Conference on Environment and Development 13, 38; Conference on the Human Environment 13, 37; Conference on Water 13; Framework Convention on Climate Change 38; Human Right to Water and Sanitation 15; Millennium Declaration 14; Millennium Development Goals 14–15, 38–39, 84; Summit on Sustainable Development 15; Sustainable Development Goals 15–16, 38–39, 84, 95–96, 139; World Water Development Report 15

urban metabolism 45, 131
urban river restoration 117
urban water sustainability 16, 174,
    178–179
Urine diverting toilets (UDTs) 93

vapour compression distillation (VCD)
    156
'virtual water' 18

washing machines 66–68, 76
wastewater 1–2, 4, 7–8, 87–91,
    131–132, 133, 139
water-based sanitation 7, 85–91, 96,
    98–99, 101, 102
water closets 85
'water crisis' discourse 47–48
'water footprint' 18
water insecurity 168

waterless sanitation 91–94
water management: integration of
    different infrastructures and sectors
    8; managing demand 6–7; paradigms
    of 3; regimes 3
water poverty 139, 161
water quality 15, 111–112
water resources 2–3, 161
water scarcity 2–3, 14, 15, 79
water sensitive cities 1, 18–20
water sensitive urban design (WSUD)
    107, 112
water use patterns 57, 59–60, 70–73,
    77
water use restrictions 64
water wastage 56, 101, 175
World Commission on Environment
    and Development (WCED) 37–38